ENVIRONMENTAL BIOGEOCHEMISTRY AND GEOMICROBIOLOGY

Volume 2: The Terrestrial Environment

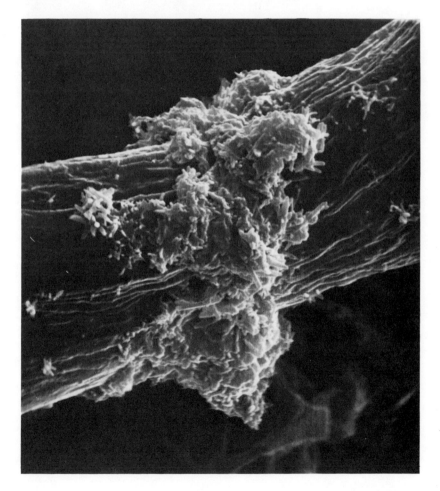

ENVIRONMENTAL BIOGEOCHEMISTRY AND GEOMICROBIOLOGY

Volume 2: The Terrestrial Environment

edited by

WOLFGANG E. KRUMBEIN
University of Oldenburg
Environmental Laboratory
Oldenburg, Germany

Proceedings of the Third International Symposium on Environmental Biogeochemistry organized by W. E. Krumbein, University of Oldenburg, and sponsored by the Minister of Science and Arts of Niedersachsen, Deutsche
Forschungsgemeinschaft, and the International Association of Geochemistry
and Cosmochemistry. The symposium was supported ideally and scientific-
ally by Deutsche Gesellschaft für Hygiene und Mikrobiologie, Gesellschaft für
Ökologie, Deutsche Bodenkundliche Gesellschaft, German Local Branch of
the American Society for Microbiology. The meeting was held at the Herzog
August Bibliothek, Wolfenbüttel.

ANN ARBOR SCIENCE
PUBLISHERS INC
P.O. BOX 1425 • ANN ARBOR, MICH. 48106

Life on earth has produced earth's present atmosphere,
the uppermost parts of the lithosphere, soil structure, water
quality and man—to adore it, study it and possibly destroy it.
Life has managed to survive long periods of dangerous fluctuations of equilibrium and, according to the "Gäa hypothesis,"
life may survive even man. Energy and entropy may be more
important to our present-day situation but without a doubt,
life and its far-reaching consequences are the most prominent
factors to be considered in studying the environment and environmental cycles.

Certainly life and its processes may be regarded as mere
chemical reactions under certain physical conditions. But
though the driving forces are primary and secondary energy
sources and flows, and though chemical balances and budgets
are possibly more basic, and though the cycle of substances
in space and time can always be expressed in chemical terms
and reactions, it must be stressed that as long as this earth
can sustain life, man will look at it and its development in
terms of a living entity. Therefore, biogeochemistry, geomicrobiology, ecology and exogenic dynamics are always regarded as events controlled and modified, speeded up or slowed
down, by life and life processes.

The production cycle for reduced carbon compounds, their
growing degree of organization, their pathway through the food
or energy chain in nature, the apparent death and final anni-
hilation by mineralization and oxidation to simple compounds—
all this can take place in a small lake or in soil within a
single day. In other circumstances, it may require billions
of years to complete a cycle. Many examples may be given for
the hypothesis that any atom that has reached or established
itself in the outer few kilometers of the earth's crust will
have passed a living organism at least once since life emerged
or, at least, will have been influenced or modified in its
energy level, spatial position, or physical-chemical condition
by organisms. It therefore seems that biogeochemistry, though
relatively far from man—at least farther than medicine, soci-
ology or music—will increasingly attract the attention of
society. Geomicrobiology and biogeochemistry will probably

increase in importance with man's growing influence on the exogenic cycle, with man's accelerating sections of the huge "mill wheel" of masses passing up and down in the earth's crust and along the energy levels.

Regarding our consternation about the carbon dioxide cycle, the importance of nitrogen compounds in the atmosphere, enormous amounts of stored energy, and reduced carbon compounds, we see the need to increase our biogeochemical activities. If we consider the acceleration of metal cycles, if we look at the enormous number of organic chemical compounds, if we take note of the changes and manipulations man produces on genetic matrices and the stress we exert on the entire ecosystem, we see that the established parameters have been drastically altered in an incredibly short time.

The aim of biogeochemistry and geomicrobiology is to find scales and balances before we so totally change our environment that no traces of the entirely "natural" remain to study, analyze and compare.

The time of holistic approaches, which science lost only 100 years ago, has returned. One major aim of this symposium series on environmental biogeochemistry is to bring together students of the various disciplines concerned with dynamic processes on the earth's surface. The main concerns are not only understanding the importance of the major mineral cycles and their budgeting and balancing, but also analyzing systems and making prognoses on these cycles that control and are controlled by life. Additional factors and study areas will include the biogeochemistry of manmade compounds and their alteration products, and the need to build up new geochemical cycles, since the natural materials used through the centuries are no longer sufficient for humanity's still-growing needs. This means that man will need to study ways to accelerate. natural mineral cycles by biological or technical methods.

PREFACE

Isotope fractionation, mobilization and immobilization, oxidation and reduction, mineralization and storage in biological material, transfer, volatilization, catalysis and equilibration of systems—all these are ruled mainly by biological processes in the microscale and most frequently by microorganisms. Therefore, microbiologists, isotope chemists, geochemists and ecologists, and a few botanists, zoologists and physicists were asked and responded to the call for papers for this meeting.

These volumes are based primarily on the papers given at the Third International Symposium on Environmental Biogeochemistry. It was our original intent to emphasize a more specific area, but the problems and the wealth of information were so broad it became necessary to expand the scope of the conference. Hence, one part of the proceedings is dedicated to defined environments, such as shallow water photosynthetic environments with dominance of blue-green algae, deep sea environments with dominance of manganese nodules, specific soil environments, or the microenvironment of one soil or rock particle, while other major segments of the proceedings are dedicated to the general cycles of elements and compounds and their alteration by man.

The study of biogeochemistry and geomicrobiology is only beginning. We wish to keep the field open and look forward to a time of organization, synthesis and perfection. The growing interdisciplinary field between biology, physical and chemical science, and geoscience cannot reasonably limit itself before the subject is more fully understood.

W. E. Krumbein

ACKNOWLEDGMENTS

The Third International Symposium on Environmental Bio-geochemistry was sponsored by Deutsche Forschungsgemeinschaft, Ministry for Science and Arts of Niedersachsen, IAGC, Balzers, Cambridge Instruments, Jürgens und Co., E. Merck, Arbeitsgemeinschaft für meerestechnisch gewinnbare Rohstoffe, Ortec, Carl Zeiss. Their help is gratefully acknowledged.

The following persons were extremely helpful in planning and encouraging the development of the meeting: E. T. Degens, K. H. Domsch, W. Flaig, P. Hirsch, M. Kürsten, G. Müller, W. Schwartz, K. Wagener and D. H. Welte of the national committee; M. Alexander, G. Eglinton, H. Ehrlich, P. H. Given, R. O. Hallberg, G. W. Hodgson, I. R. Kaplan, K. A. Kvenvolden, A. D. McLaren, P. A. Meyers, J. O. Nriagu, E. A. Paul, M. Schnitzer, J. Skujinš and G. Stotzky of the international committee, and namely J. Skujinš, chairman of the IC, K. Kvenvolden and E. Ingerson of IAGC, J. W. M. la Rivière of SCOPE, and A. Meyl of DFG.

It is my pleasure to acknowledge the support of P. Raabe of the Herzog August Bibliothek, Wolfenbüttel, and his staff. Without their aid and acceptance, and especially the calm and efficient stability of D. E. Petersen, the meeting would not have been possible. I also wish to extend my thanks to W. Schwartz who was so kind to cooperate and coordinate the "Roundtable Conference on Leaching" with this symposium.

Christine Lange, Peter Rongen, Elisabeth Holtkamp, Joachim Leibacher, Cornelia Wilcken, Monika Michaelsen, Ulrike Kant and G. Koch have made a major contribution to the success of the practical arrangements and to the well-being of the participants.

Finally I wish to express my gratitude to all participants for attending and submitting their contributions on schedule.

W. E. Krumbein, Associate Professor at West Germany's University of Oldenburg, is the chairman and coordinator of the Third International Symposium on Environmental Biogeochemistry. An expert on microbiological rock weathering, his work has concentrated on geomicrobiology, environmental biogeochemistry, productivity and element cycles in natural environments.

Dr. Krumbein received his Vordiplom (BSc) in geoscience from the University of München, and his MSc and PhD (both magna cum laude) from the University of Würzburg. He studied soil microbiology and microbial ecology at the Institut Pasteur and the Sorbonne in Paris and the Landgebouwhoogeschool Wageningen in the Netherlands, and was the recipient of a research grant for postgraduate study at Jerusalem's Hebrew University. He then spent six years as a research scientist at the Biologische Anstalt Helgoland, a government laboratory for marine research.

The editor has taught on the faculties of the Universities of Würzburg, Freiburg and Hamburg. He was a guest scientist at the Scripps Oceanographic Institute in La Jolla, California, and was recently a Visiting Professor at Hebrew University.

Dr. Krumbein has been a speaker at many international scientific conferences, was active as Convener of the International Congresses of Sedimentology and Ecology, and has completed three lecture tours in the U.S. For his active promotion of the fields of geomicrobiology and biogeochemistry, Dr. Krumbein has received awards from the Deutsche Geol. Ges. (German Geological Society) and the Institut Pasteur. He is a member of the Advisory Board of the MBL, Elat, ISEB's International Committee, the National Building Research Council and ICES/ECOMOS. He is also Associate Editor of *Geomicrobiology Journal*.

At present, Dr. Krumbein and his colleagues are planning to establish an institute for salt water biology in Germany. In recent papers he has dealt with hypersaline environments and calcification in prokaryotic organisms as well as with nitrogen and phosphorus budgets of intertidal cyanobacterial communities of the Gulf of Aqaba and the North Sea coast.

TABLE OF CONTENTS

SECTION III
SHALLOW PHOTIC WATERS AND SEDIMENTS WITH SPECIAL REFERENCE
TO STROMATOLITIC ENVIRONMENTS

SECTION IV
DIAGENESIS

VOLUME 2

SECTION I
GEOMICROBIOLOGY, BIOGEOCHEMISTRY AND ENERGY FLOW

SECTION II
PADDY SOILS, PEAT AND COAL

SECTION III
NITROGEN IN SOIL AND ITS IMPACT ON THE ATMOSPHERE

SECTION IV
INTERFACES AND SORPTION

SECTION V
DESTRUCTION, MINERALYSIS, WEATHERING

VOLUME 3

SECTION I
METHODS TO ASSESS BIOGEOCHEMISTRY AND GEOMICROBIOLOGY OF THE ENVIRONMENT

SECTION IV
BIOGEOCHEMISTRY OF MAN'S FINGERPRINTS IN NATURE
(METAL-ORGANIC RELATIONS)

SECTION I

GEOMICROBIOLOGY, BIOGEOCHEMISTRY AND ENERGY FLOW

CHAPTER 31

MICROFLORA OF LAND MICROENVIRONMENTS
AND ITS ROLE IN THE TURNOVER OF SUBSTANCES

G. I. KARAVAIKO

Institute of Microbiology
USSR Academy of Sciences
Moscow, USSR

Land biogeochemical processes involve in fact only the zone of hypergenesis, which is a complex system of various microenvironments or natural bodies. To evaluate the activity of microorganisms in the zone of hypergenesis now, we considered the following land microenvironments: (1) rocks, (2) igneous and metamorphic rocks of weathering crusts, (3) ore deposits, and (4) organogenous sediments.

Our approach to the understanding of land microenvironments is based on estimating them as a specific habitat of microbial cenoses. This is what constitutes relative differences between various land microenvironments, which are also interrelated by common processes of transformation and migration of elements and substances.

MICROFLORA OF ROCKS

Igneous and metamorphic rocks are generally believed to be a poor environment for the growth of microorganisms. However, many authors found diverse and numerous microflora in the products of rock degradation (Krasilnikov, 1949; Glazovskaja, 1950a,b; 1952; Omelyansky, 1953; Polynov, 1953; Webley *et al.*, 1963; Vernadsky, 1960; Zajic, 1967; Schwartz and Schwartz, 1972). The cenoses of microorganisms and processes were found to be similar, regardless of geographical zones and rock types. Algae and oligotrophous nonsporeforming bacteria belonging to the genera *Pseudomonas*, *Corynebacterium* and *Arthrobacter* predominate in the rocks. Many of them are capable of fixing atmospheric nitrogen. The cenoses include the most primitive, blue-green algae (*Chroococcales*) which are not encountered in most soils. Rocks disintegrate mainly along cracks and at the surface. Accumulation of organic

substances, leaching of elements (S, Ca, K, Fe), and formation
of clayey minerals occur in weathering crusts.

Crusts of weathering, which are accompanied by large
deposits of bauxites, kaoline clays and nickel, are more complex land microenvironments. Their formation is related to
the degradation of original rocks under certain oxidative-
reductive conditions. Some data about the microflora of
weathering crusts are presented in Table I. First of all,
weathering crusts are characterized by general distribution
of autotrophous bacteria, which oxidize reduced compounds of
nitrogen, sulfur and iron. Nonsporeforming bacteria prevail
among heterotrophs, particularly, as we have shown, those belonging to the genera *Arthrobacter, Corynebacterium* and
Mycobacterium. In a pure culture, they assimilate phenols,
humates, ethanol and other alcohols, and grow on nitrogen-poor
rocks. Many of them are capable of fixing atmospheric nitrogen.

The microflora of sedimentary rocks in the zone of hypergenesis is more diverse, and the geochemical activity of
various bacterial groups depends on the reduction-oxidation
conditions of the medium.

As can be seen in Figure 1, zones originate in urani-
ferous rocks of the Mesozoic-Cenozoic deposits along the flow
of underground waters. Here, an oxidation zone of limonitization and leaching of elements is followed with a reduction
zone or a zone of secondary uranium enrichment, and then with
a zone of unchanged rocks (Lisitzyn and Kuznetzova, 1967).
The most diverse and active microflora is found at the boundary
between the oxidation and reduction zones: thiobacilli, nitrifying and denitrifying bacteria, microorganisms oxidizing H_2
and hydrocarbons, and producing H_2, H_2S, and CH_4. Under
reductive conditions, sulfate-reducing, hydrogen- and methane-
forming bacteria predominate; they create a geochemical barrier
for precipitating uranium, selenium, and other elements from
underground waters.

As follows from these studies, microorganisms are involved not only in the leaching and migration of elements, but
also in the formation of minerals under present conditions.
The participation of microorganisms in hypergenic processes in
rocks is schematically presented in Figure 2.

While evaluating the conditions of microbial growth
in rocks, one has to find the sources of carbon and energy.
According to geochemical data, metal sulfides are widely distributed among both the original metamorphic rocks of weathering crusts and sedimentary rocks (Nadezhdina, 1961; Lyakhovich,
1964; Andreyuk *et al.*, 1976). The content of sulfides,
especially of pyrite, varies from hundredth fractions of gram
to 2-40 kg/m^3 of rock. Active acidity in rocks often rises
sharply (pH = 1.4-3.0) due to bacterial oxidation of pyrite
and sulfur. Reduced sulfur compounds are also formed as a
result of the viability of sulfate-reducing bacteria.

Table I

Microorganisms of Crusts of Weathering[a]

Weathering Crust	Biogeochemical Processes	Microorganisms
Weathering crust of Mesozoic-Cenozoic deposits (shales) of the Western Siberian Platform (Tomsk)	Oxidation of organic substances and pyrite; leaching of elements; formation of clayey minerals (pH 1.4–8.0)	*T.ferrooxidans*, *T.thiooxidans*, thiobacilli oxidizing $S_2O_3^{2-}$, heterotrophous (mainly nonspore-forming) organisms
Weathering crust of ultra-basic rocks of the Urals and Cuba	Oxidation of NH_4^+, NO_2^-, pyrite, organic substances; degradation of serpentines yielding nontronite and ocher; leaching of Mg, Ni, Si, Ca and other elements.	Nitrifying bacteria (I and II phases), *Thiobacillus* sp.; heterotrophous microorganisms
Degradation of pegmatite and pyroxene in spodumene deposits (Eastern Siberia)	Oxidation of organic substances and reduced sulfur compounds; degradation of pegmatite, spodumene and shale; leaching of Li, Al, Si, Fe; formation of kaolinite and montmorillonite (pH 2.8–7.2)	*Mycobacterium*, *Arthrobacter*, *Corynebacterium*, *Pseudomonas*, *Azotobacter*, nitrifying bacteria (I and II phases), *Desulfotomaculum*, *T.ferrooxidans*, *T.thiooxidans*, thiobacilli similar to *T.thioparus*, fungi

[a]Udodov *et al.*, 1974; Lebedeva *et al.*, 1977; our data.

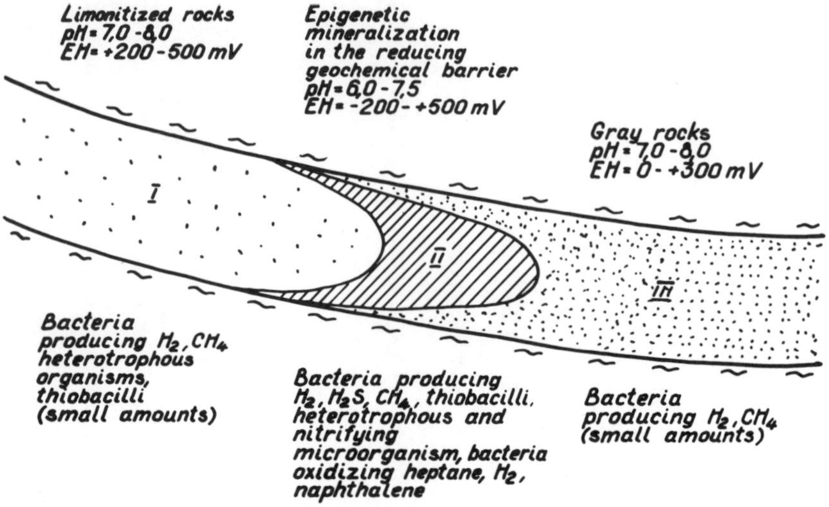

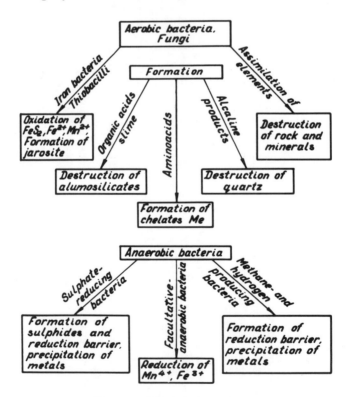

Figure 1. Changes in the environmental conditions and in the composition of microflora in the direction of flow of stratum waters in gray rocks of the uplift of the artesian basic.

Figure 2. Role of microorganisms in destruction of rocks.

Sources of energy and carbon for heterotrophs may be also dispersed organic substances up to 20 mg/kg of bitumen, and some gaseous products (H_2 up to 630 cm^3/kg, CH_4 up to 100 cm^3/kg and CO_2) of rocks and precipitation (up to 20 mg/l simple organic substances) (Semenov *et al.*, 1966; Petersilye and Pavlova, 1975; Karzhavin, 1975; Pavlova, 1975). Pore solutions of sedimentary rocks contain dozens and hundreds of milligrams of organic carbon per liter (up to 35.5 mg of volatile organic acids and 20 mg of naphthenic acids per liter) (Kartzev *et al.*, 1970). Belyayev (1976) found 0.6-0.8 mg of acetic acid per liter of underground waters in the pre-Carpathian sulfur deposits. It is generally accepted now that oligotrophous microorganisms can grow at the expense of minimum amounts of organic substances in waters and rocks: 0.2 mg of organic carbon per liter, 0-100 mcg of phenol per liter, 0.1 percent of bitumen (Ivanov, 1964; Lapteva, 1975; Romanenko and Kuznetzov, 1976). Simple organic compounds are assimilated by microorganisms at a very low concentration, 5 mcg/l (Wright and Hobbie, 1965; Allen, 1969).

The second source of organic matter for heterotrophs in rocks is metabolites of photo- and chemolithotrophic bacteria. Algae and thiobacilli are known to liberate 5-50 and 20-50 percent of fixed carbon, respectively, as organic compounds into the medium (Fogg *et al.*, 1965; Karavaiko and Pivovarova, 1973; Kuenen, 1975). A possibility of growth and biomass accumulation of heterotrophs in cultures of *Thiobacilli* and nitrifying bacteria has been proven experimentally (Karavaiko *et al.*, 1973; Golovacheva, 1976). Therefore, it is obvious that rocks contain organic substances at concentrations which considerably exceed those required for the beginning of growth of microbial cenoses. Certain species of hydrogen bacteria can fix nitrogen and grow autotrophically (Gogotov and Schlegel, 1974; Wiegel and Schlegel, 1976). This ability is manifested also by certain species of *Corynebacterium* that we found in rocks. It is quite possible that they may be capable of autotrophous growth under natural conditions in rocks in the presence of H_2, CH_4 and CO_2, but this must be proven experimentally.

The trophic relationships of different microbial groups in various geochemical crusts of weathering may be represented, according to the evidence available, as a scheme in Figure 3.

MICROFLORA OF ORE DEPOSITS

Ore deposits are characterized by a considerable concentration of elements, indicating that biogeochemical processes of their transformation proceed on the geological scale. Of course, ore deposits are an excellent ecological niche for the activity of a specific autotrophic microflora. *Thiobacilli* utilizing the energy of oxidation of Fe^{2+}, sulfur and its reduced compounds predominate in the deposits of sulfur and

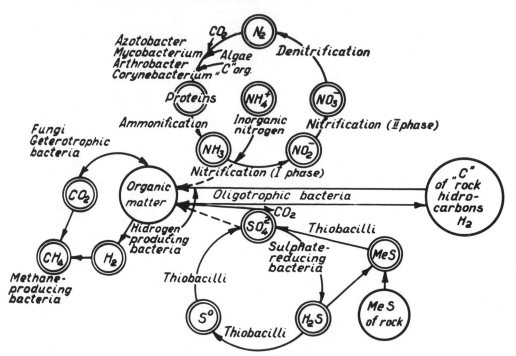

Figure 3.   Scheme of the trophic relationship between microorganisms in rocks.

sulfide ores (Sokolova and Karavaiko, 1964; Zajic, 1967;
Karavaiko *et al.*, 1972).

A number of new microorganisms oxidizing various elements
were found in ore deposits: *Leptospirillum ferrooxidans*
($Fe^{2+} \longrightarrow Fe^{3+}$), *Thiobacillus organoparus* and *Thiobacillus
acidophilus* ($S^{o} \longrightarrow SO_4{}^{2-}$) (Markosyan, 1972, 1973; Guay and
Silver, 1975), *Stibiobacter senarmontii* ($Sb^{3+} \longrightarrow Sb^{5+}$)
(Lyalikova, 1974).   *Thiobacillus "y"* oxidizes several sulfide
minerals (Sb_2S_3, PbS, Bi_2S_3) at high pH values (pH = 7.5–8.0)
of the medium (Lyalikova, 1967; Lyalikova and Kulikova, 1969).
A mixed bacterial culture oxidizing MoO_2 has been isolated
(Lebedeva and Lyalikova, 1975).   Apparently, some specific
bacteria can also oxidize such elements as Co^{3+} and As^{3+}.
Under favorable physicochemical conditions, elements are
leached from ores and transferred with ore waters.   Leaching
often proceeds on such a large scale that it can be a source
for obtaining considerable amounts of metals.

MICROFLORA OF ORGANIC SEDIMENTS

Organic sediments are a land microenvironment, which
allows observation of how biogeochemical transformation of
complex organic substances changes with physicochemical

conditions of the medium. The formation of biogeochemical zones and the character of a microbial cenosis depend on the water exchange of strata (Table II). The hydrocarbon-oxidizing microflora of oil deposits is rather uniform and represented by the genera *Pseudomonas* and *Mycobacterium*. The former genus is predominant because it is more adapted to assimilate liquid n-alkanes (C_5-C_{10}) having a low boiling point and aromatic hydrocarbons, and it can grow under microaerophilic and anaerobic conditions (Rozanova and Kuznetzov, 1974).

Microbial cenoses include also methane- and hydrogen-producing microorganisms, thiobacilli, and sulfate-reducing bacteria. Thermophilic sulfate-reducing bacteria were found at a depth of 4000 m (Rozanova, 1975). Nonhydrocarbon components of oil are also decomposed by microorganisms. *Mycobacterium* and *Brevibacterium* are involved in the oxidation of ozocerite (Rozanova and Kuznetzov, 1974). Microorganisms belonging to the genera *Pseudomonas, Achromobacter, Micrococcus, Flavobacterium, Bacillus, Mycobacterium,* and *Nocardia* may also destroy asphaltum (Traxler *et al.*, 1965, 1966).

Microbiological degradation of oil in deposits proceeds on a very large scale. According to Williams *et al.* (1972), 10 percent of oil deposits in the world are destroyed by microorganisms, and the quality of an additional 10 percent sharply decreases.

The rate of microbiological processes in oil deposits increases as the water exchange in strata becomes more active as a result of oil extraction, particularly, the rate of hydrocarbon degradation and sulfate reduction. The following practical problems arise in this case (Rozanova, 1975): the prevention of corrosion of the equipment caused by hydrogen sulfide, an increase in the output of oil by the stratum, production of biogenic hydrogen sulfide in exhausted oil deposits, desulfuration of oil and production of microbial protein from oxidized organic substances of exhausted oil deposits.

Sedimentary deposits with dispersed organic substances are characterized by microbiological generation and oxidation of hydrocarbon gases and organic matter. Sulfate reduction is also often encountered and is particularly active in gas-oil regions where underground waters are enriched with organic substances, and hydrocarbon gases migrate from oil-bearing deposits. *Mycobacterium* and *Pseudomonas* predominate among microorganisms oxidizing hydrocarbon gases in sedimentary strata (Telegina, 1965), and the oxidation is specific in various microorganisms that are used in detecting gas-oil deposits. Bacteria that oxidize propane, butane and, partly, pentane are most characteristic. The effectiveness of geomicrobiological methods employed in complex with other direct methods is 60-70 % (Mogilevsky, 1975).

Table II

Biogeochemical Zones of Sedimentary Deposits and Their Characteristics[a]

Biogeochemical zones, Mineralization of Water	Biocenosis of Microorganisms	Manifestation of Biogenic Effect
I. Surface deposits enriched with ozocerite; surface precipitation	Hydrocarbon oxidizing (*Mycobacteriaceae*, *Brevibacteriaceae*, *Candida*, fungi)	Harsan kyrocks; depletion of paraffin from ozocerite, which becomes more solid and resinaceous
II. Active water exchange; diffusion of hydrocarbon gases, 0.05-1 g/l	Hydrocarbon oxidizing (*Pseudomonadaceae*, *Mycobacteriaceae*)	Microorganisms oxidizing gaseous hydrocarbons
Active water exchange; enrichment of waters with oil bituminous matter, 0.5-2 g/l	Hydrocarbon oxidizing (*Pseudomonadaceae*, sulfate-reducing, methane-forming, *thiobacilli*)	Maltha; tarry oil with a low content of paraffins; water with a high content of oxidized organic matter, H_2S and CO_2, and a low content of H_2SO_4; karst is formed on gypsum
III. Hampered water exchange; enrichment of waters with oil bituminous matter, 20-200 g/l	Sulfate-reducing bacteria predominate	H_2S in oil and water; the content of CO_2 and NH_3
IV. Water exchange is very slow or absent; enrichment of waters with oil, over 200 g/l	Absent	Relict; H_2S in oil and water; the content of CO_2 and NH_3 in water increases; formation of secondary calcite impregnated with black oil in the zone of oil-water contact

[a]according to E. P. Rozanova.

Mineral coal is a mixture comprised mostly of organic and partly of mineral substances. The basic organic matter of coal consists of neutral humic compounds and a small amount of bitumen. Brown coal is an intermediate stage between coal and peat, and contains up to 60 percent of humic acids. The content of sulfur (iron sulfides) in different types of coal varies from fractions of a percent to 6-7 percent. Coals also contain ions of metals: Fe^{2+}, Fe^{3+}, Co^{2+}, Zn^{2+}, Mo^{2+}, Be^{2+}, Ni^{2+}, and Sr^{2+} which pass to a solution and are leached in the zone of oxidation.

The microflora of coal deposits hitherto has not been studied sufficiently well. The main information about the geochemical activity of microorganisms in coal-bearing deposits is presented in the review by Nesterov (1976). This microflora is specific and consists by 80 percent of *Pseudomonas* bacteria. The microflora includes also microscopic fungi, actinomycetes, hydrogen-producing, methane-forming and sulfate-reducing bacteria. The bacteria isolated from coals are mainly facultative or strict anaerobes. *Thiobacilli* (*T.ferrooxidans*, *T.thiooxidans*) and methane-oxidizing bacteria are widely distributed (10^3-10^7 cells/ml) in the zone of oxidation.

Oxidation processes become more active in coal deposits in the course of their exploitation when coal is crushed and wetted. The basic microbiological processes in coal pits may be represented by a scheme shown in Figure 4 (according to Nesterov, 1976).

Studies of the microflora of coal deposits are important not only in order to elucidate the genesis of coal and gases, but also to eliminate methane by oxidizing it with the aid of bacteria, to remove sulfur from coal, and to suppress the activity of *thiobacilli* that produce aggressive acidic pit waters. However, a method of bacterial oxidation of methane in coal strata has been the only one tested hitherto in the USSR (Moskalenko *et al.*, 1976). Its effectiveness reaches 60 percent.

Analysis of literature on the microflora of rocks and organic sediments makes it possible to draw the following conclusions. Biogeochemical processes in rocks and crusts of weathering are determined by cenoses of microorganisms that are in certain trophic relationships. The concentration of organic substances and gases in rocks seems to be sufficient to start the growth of microorganisms.

Autotrophic microorganisms of rocks and their geochemical activity are an important object of studies as sources of energy and carbon, particularly oligotrophic bacteria and corynebacteria, fixing nitrogen and assimilating H_2 and CO_2, respectively. According to the data of geochemistry, rocks contain not only dispersed organic substances but also gases

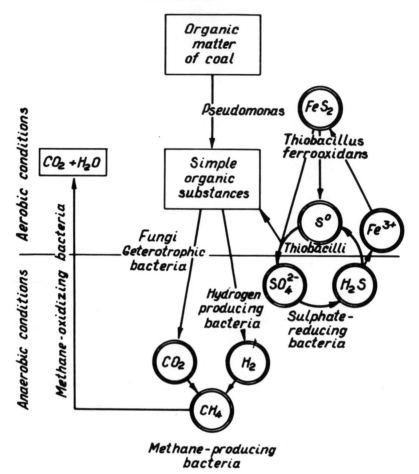

Figure 4.   Scheme of microbiological processes in coal
deposits.

(H_2, CO_2, CH_4).   Under certain conditions, these gases are
formed by microorganisms in specific land microenvironments
(ore deposits, organic sediments).   The character and scale
of biogeochemical processes are determined by specific autotrophic or heterotrophic microorganisms.   Microbial cenoses
are similar in respective land microenvironments of different
climatic zones.   It is still difficult to estimate the scale
of the geochemical activity of microorganisms in land microenvironments because enough quantitative data are not available.
However, it is safe to state that microorganisms play a leading role in the turnover of elements in the zones of hypergenesis.   These processes become more intensive as a result
of man's economic activity.   These processes may be either
useful (leaching of metals, concentration of elements) or
harmful (corrosion of equipment, deterioration of ore tech-
nological qualities).

REFERENCES

Allen, H. L. "Chemoorganotrophic Utilization of Dissolved Organic Compounds by Planktic Algae and Bacteria in a Pond," *Int. Revue Ges. Hydrobiol.* 54(1):1-33 (1969).

Andreyuk, E. I., L. I. Rubenchik, I. A. Kozlova and N. S. Antonovskaya. "Thiobacilli as a Factor-Producing Aggressive Medium in Underground Constructions," in *Ecology and Geochemical Activity of Microorganisms*, M. V. Ivanov, Ed., (Moscow: Pushchino Publications, 1976), pp. 135-139.

Beliayev, S. S. "Geochemical Activity of Methane-Forming Bacteria," in *Ecology and Geochemical Activity of Microorganisms*, M. V. Ivanov, Ed., (Moscow: Pushchine Publications, 1976), pp. 139-152.

Fogg, G. E. and W. D. Watt. "The Kinetics of Release of Extracellular Products of Photosynthesis by Phytoplankton," *Mem. Ist. Ital. Hidrobiol.* 18 (suppl.):165 (1965).

Glazovskaya, M. A. "The Effect of Microorganisms on Weathering of Primary Minerals," *Izv. AN Kaz. SSR, soil ser.* 86(6): 79-100 (1950a).

Glazovskaya, M. A. "Weathering of Rocks in the Nival Belt of the Central Tien Shan," *Works of V. Dokuchayev Soil Inst.* 34:28-48 (1950b).

Glazovskaya, M. A. "Loose Products of Rock Weathering and Primary Soild in the Nival Belt of the Terskey-Alatau Ridge," *Works of Geogr. Inst. of USSR Acad. Sci.* 49(2): 70-129 (1952).

Gogotov, J. N. and H. G. Schlegel. "N_2-Fixation by Chemoautotrophic Hydrogen Bacteria," *Arch. Microbiol.* 97(4): 359-362 (1974).

Golovacheva, R. S. "Thermophilic Nitrifying Bacteria of Hot Springs," *Mikrobiol.* 45(2):377-379 (1976).

Guay, R. and M. Silver. "Thiobacillus Acidophilus Sp. Nov., Isolation and Some Physiological Characteristics," *Canad. J. Microbiol.* 21(3):281-289.

Ivanov, M. V. *The Role of Microbiological Processes in Genesis of Sulphur Deposits* (Moscow, Nuaka Publications 1964).

Karavaiko, G. I. and T. A. Pivovarova. "Oxidation of Elemental Sulphur by Thiobacillus Thiooxidans," *Mikrobiol.* 42(3):389-395.

Karavaiko, G. I., S. I. Kuznetzov and A. I. Golomzik. *The Role of Microorganisms in Metal Leaching from Ores* (Moscow, Nauka Publications, 1972).

Karavaiko, G. I., E. V. Shchetinina, T. A. Pivovarova and K. Yu. Mubarakova. "Denitrifying Bacteria Isolated from Sulphide Ore Deposits," *Mikrobiol.* 42(1):128-135.

Karzhavin, V. K. "Studies of Products of Thermodesorption of Apatite and Nepheline and Kinetics of Gas Evolution," in *Carbon and Its Compounds in Endogenous Processes of Mineral Formation* (Lvov Publications, 1975), pp. 31-32.

Kartzev, A. A., M. Ya. Dudova and O. P. Abramova. "Organic Matter in Rock Solutions," *Izv. AN SSSR, Geol. Ser.* 7: 137-138.

Krasilnikov, N. A. "The Role of Microorganisms in Rock Weathering, Microflora of the Surface Layer of Rocks, *Mikrobiol.* 18(4):318-323.

Kuenen, J. G. "Colorless Sulfur Bacteria and Their Role in the Sulfur Cycle," *Plant a. Soil.* 43:49 (1976).

Lapteva, M. A. "Growth of Bacteria on the Media with Minimum Content of Organic Substances," *Vth Conf. All-Union Microbiol. Soc.* pp. 58-59.

Lebedeva, E. V. and N. N. Lyalikova. "Geochemical Role of Bacteria in Molybdenum Oxidation in the Tyrnyauz Molybdenum-Tungsten Deposit," *Vth Conf. All-Union Microbiol. Soc.* pp. 34-35.

Lebedeva, E. B., N. N. Lyalikova, Yu. Yu. Bugelsky and O. A. Maslova. "Participation of Nitrifying Bacteria in Weathering of Serpentized Ultrabasic Rocks," *Mikrobiol.* 46(3).

Lisitzyn, A. K. and E. G. Kuznetzova. "The Role of Micro-Organisms in the Formation of Reducing Geochemical Barriers in the Zones of Stratum Limonitization," *Izv. AN SSSR, Geol. Ser.* 1:31-43.

Lyalikova, M. N. "Oxidation of Antimonite by a New Culture of Thiobacilli," *Dokl. AN SSSR* 176(6):1432-1434.

Lyalikova, N. N. "Stibiobacter Senarmontii, A New Micro-Organism Oxidizing Antimony," *Mikrobiol.* 43(6):941-948.

Lyalikova, N. N. and M. F. Kulikova. "Oxidation of Germanium-Containing Sulphides by Bacteria," *Dokl. AN SSSR* 184(5): 1194-1196.

Lyakhovich, V. V. *Accessory Minerals of Granitoids* (Moscow, IMGRE Publications).

Markosyan, G. E. "A New Iron-Oxidizing Bacterium, Leptospirillum Ferrooxidans Nov. Gen., Nov. Sp.," *Biol. J. Armen.* 25(2):26-29.

Markosyan, G. E. "A New Myxotrophous Species of Thiobacillus Organoparus Growing in Acid Medium," *Dokl. AN SSSR* 211(5):1205-1208.

Mogilevsky, G. A. "Geomicrobiological Studies of Oil, Gas and Coal Deposits Now and in the Future," *Mikrobiol. Prom.* 1:3-6.

Moskalenko, E. M., M. V. Ivanov, A. I. Nesterov, N. G. Smolyaninov, B. N. Perminov and A. V. Nazarenko. *Microbiological Oxidation of Methane in Coal*, M. V. Ivanov, Ed., (Moscow, Pushchino Publications, 1976), pp. 172-177.

Nadezhdina, E. D. *Accessory Minerals of Traps near the Podkamennaya Tunguska River*, A. P. Lebedev, Ed., (Moscow, IGEM Works, USSR Acad. Sci. Publications, 1961).

Nesterov, A. I. "Microflora of Coal-Bearing Deposits and Its Geochemical Activity," in *Ecology and Geochemical Activity of Microorganisms*, M. V. Ivanov, Ed., (Moscow: Pushchino Publications, 1976), pp. 164-172.

Omelyansky, V. L. "The Role of Microorganisms in Rock Weathering," in *Selected Works*, A. A. Imshenetsky, Ed., (Moscow, USSR Acad. Sci. Publications, 1953), pp. 523-530.

Pavlova, M. A. "Experimental Studies of Thermic Transformation of Bitumen Substances in Igneous Rocks of the Khibini Massif," in *Carbon and Its Compounds in Endogenous Processes of Mineral Formation,* (Lvov Publications, 1975), pp. 102-104.

Petersilye, I. A. and M. A. Pavlova. *Organic Compounds in Igneous Rocks*, (Lvov Publications, 1975), pp. 28-30.

Polynov, B. B. "The Geological Role of Organisms," in
   Problems of Geography, M. A. Glazovskaya and N. A.
   Gvozdetzky, Eds., (Moscow, Geogr. Lit. Publications, 1953),
   pp. 45–64.

Romanenko. V. I. and S. I. Kuznetzov. "Utilization of Labeled
   ^{14}C-Phenol for Determination of Its Reserves and the Rate
   of Assimilation by Microflora of Reservoirs," *Mikrobiol.*
   45(6):166–168.

Rozanova, E. P. "Microbioligical Studies Related to Exploitation of Oil Deposits," *Mikrobial Prom.* 1:33–37.

Roznova, E. P. and S. I. Kuznetzov. *Microflora of Oil
   Deposits,* (Moscow, Nauka Publications, 1974).

Savostin, P. "Microbial Transformation of Silicates,"
   Zeitschr. F. Pflanzenernahrung U. Bodenkunde 132(H.1):37–45.

Schwartz, W. and A. Schwartz. "Geomikrobiologische
   Untersuchungen. X. Besiedelung der Vulkaninsel Surtsey mit
   Mikroorganismen," *Zeitschr. Allg. Mikrobiologie* 12(4):287–300.

Semenov, A. D., L. I. Nemtsova, T. S. Kishkinova,and A. P.
   Pashanova. "On Content of Individual Groups of Organic
   Substances in Precipitation," in *Hydrochemical Materials*
   42:17–21.

Sokolova, G. A. and G. I. Karavaiko. *Physiology and
   Geochemical Activity of Thiobacilli,* S. I. Kuznetzov, Ed.
   (Moscow, Nauka Publications, 1964).

Telegina, Z. P. "Distribution of Microorganisms Assimilating
   Gaseous Hydrocarbons and their Utilization for Detecting
   Oil," *Cand. Sci. Diss. Theses* (Moscow, 1968).

Traxler, R. W., P. R. Proteau and R. N. Traxler. "Action
   of Microorganisms on Bituminous Materials, I. Effect of
   Bacteria on Asphalt Viscosity," *Appl. Microbiol.*
   13:838–841.

Traxler, R. W., J. A. Robinson, D. E. Wetmore and R. N.
   Traxler. "Action of Microorganisms on Bituminous Materials,
   II. Composition of Low Molecular Weight Asphaltic Fractions
   Determined by Microbial Action and Infrared Analyses,"
   J. Appl. Chem. 16:266–271.

Udodov, P. A., E. P. Shamolina, E. S. Korobeinikova and N. A. Korovina. "Studies of Geochemical Activity of Thiobacilli Isolated from Pore Solutions," in *Problems of Geochemistry of Underground Waters in Respect to Detection of Mineral Ores,* (Tomsk Publications, 1974), pp. 102–108.

Vernadsky, V. I. "On Kaoline Degradation by Organisms," in *Problems of Geochemistry of Underground Waters in Respect to Detection of Mineral Ores,* (Tomsk Publications, 1960), 5:118–119.

Webley, D. M., M. E. K. Henderson and I. F. Taylor. "The Microbiology of Rocks and Weathered Stones," *J. Soil. Sci.* 14(1):102–112.

Wiegel, J. and H. G. Schlegel. "Enrichment and Isolation of Nitrogen Fixing Hydrogen Bacteria," *Arch. Microbiol.* 107(2):139–142.

Williams, I., R. Harwood, W. Dow and I. Winters. *World Oil* 1:28–29.

Wright, R. T. and J. E. Hobbie. "The Uptake of Organic Solutes in Lake Water," *Limnol. A. Oceanogr.* 10:(1):22–28.

Zajic, J. E. *Microbial Biogeochemistry* (New York, London Acad. Press, 1967).

CHAPTER 32

ENERGY FLUXES IN SOIL MICROENVIRONMENTS

AMYAN MACFADYEN

New University of Ulster
Coleraine
Londerry, Northern Ireland

In some soils organic matter is thought to be broken down directly by chemical action (*e.g.*, Swaby and Passey, 1953) whereas in others organic matter is not broken down over very long periods but accumulates, as peat or coal for instance. But in areas that are familiar in temperate regions and exploited by man, only a small accumulation of organic matter occurs. Recent findings (*e.g.*, Jenkinson, 1971, Stout *et al.*, 1976), do demonstrate however, that a small proportion becomes highly resistant to decay and may have a half-life of 1000 years or more.

In such soils the ultimate source of organic matter is the photosynthetic activity of plants, which do not usually capture more than 1 or 2 percent of the radiation reaching the earth's surface. The way in which photosynthetic energy, bound by plants in high energy carbon compounds, is shared out thereafter is demonstrated in Figure 1, which represents two fairly extreme cases, a grazed grassland and a beechwood forest (Macfadyen, 1970). It should be noticed that because trees carry much more living matter per unit area, the proportion of gross production (photosynthesis) used in vital processes (respiration) by the trees is much higher than in the grasses. However, the gross production itself is also higher and the proportion eaten directly from the plant by herbivores is lower. As a result the amount that passes to the soil is much the same, and it is a useful rule of thumb that about 10 MJ m^{-2} yr^{-1} is the energy input in the form of litter passing to most familiar terrestrial systems (excluding, of course, those whose productivity is artificially enhanced by agricultural practices).

There are several other sources of energy to the soil as demonstrated in Table I, especially dead roots (whose physiology and productivity is still not well studied), which appear in some cases, such as prairie grassland (Coleman,

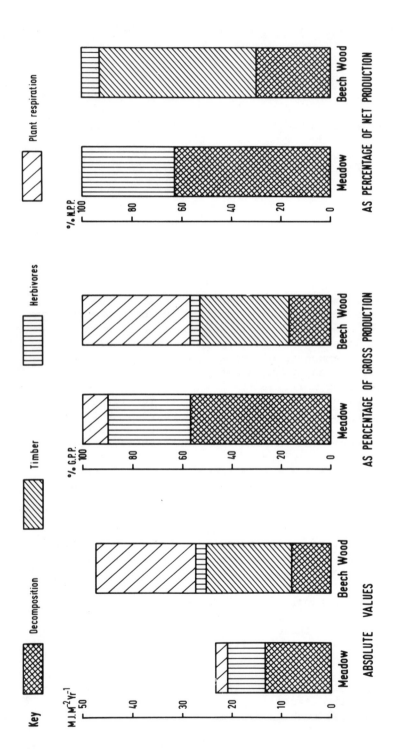

Figure 1. To demonstrate the share-out of photosynthetic energy in two representative temperate terrestrial ecosystems, a grazed meadow and a beechwood forest (Macfadyen, 1970) are compared. Note that high plant respiration in woodland reduces the proportion of gross production passing to decomposition but the smaller proportion consumed by herbivores restores the proportion of net production. The energy in both systems passing to decomposers is close to the widely relevant figure of 10 MJ m^{-2} yr^{-1}.

Table I
Approximate Quantities, Sources and Losses
of Carbon and Energy as MJ m^{-2} yr^{-1} and kg CO_2 m^{-2} yr^{-1}
Associated with the Supply of 0.5 kg m^{-2} yr^{-1} Litter

Energy Sources	Outputs of Carbon Dioxide
Litter[a] 10 MJ	14.0 kg
Dead Roots[b] 2	1.4
Root Exudates[b] 2	1.4
Mycorrhiza[c]	0.7
Total	17.5
Plus Root Respiration	4.4
Grand Total	21.9

[a]Macfadyen, 1970.
[b]Coleman, 1973,1976.
[c]Hartley, 1973.

1973; 1976) to contribute as much as above-ground production. This contribution tends to take place later in the growing season than above-ground growth. In addition, plant roots lead soluble compounds such as sugars and amino acids, which can contribute substantially to soil energy supplies. Finally the symbiotic root-inhabiting fungi called Mycorrhiza derive much energy from roots, and ultimately liberate this in the soil (Hartley, 1973).

However, the above ground litter supply normally constitutes about two-thirds of the total. In broad terms our figure of 10 MJ m^{-2} yr^{-1} can stand as an indicator of the order of magnitude of the total input. One method of measuring total biological soil energy flux is to treat the soil as a combustion system and measure the carbon dioxide output. The biggest problem here, however, is that a large proportion of the carbon dioxide, a proportion which in many soils is still very uncertain, is derived from the respiration of the roots themselves.

The figure of 10 MJ yr^{-1} represents only about 0.32 W continuous energy liberation and is therefore trivial in comparison with the order of 170 W, which is a typical short-wave energy gain in temperate latitudes (Macfadyen, 1967). Nevertheless the biologist is concerned with investigating the biologically determined energy flux further for two main reasons. First, energy-liberating activities, although relatively low, are highly localized both in space

and in time. Spatially, the soil habitat is extremely complex, containing air and water, as well as organic and inorganic particles. The situations suitable for growth are
very limited and even suitable regions are quite sparsely
colonized. As regards distribution of activity in time, it
is clear from calculations of the difference between potential growth rates of microorganisms (Gray, 1976) and their
actual performance in soil, that the majority are almost
inert for most of the time but are capable of brief periods
of intense activity. It is these features of heterogeneity
that have frustrated the development of biological studies
in soil more than any others.

The second reason for concern with biological energy flux
is that energy liberation is inextricably linked with the
release from organic compounds of simple compounds and radicles
containing nitrogen, phosphorus and potassium, especially.
If these are not returned to the plant roots, the plants –
and in turn the remaining organisms in any habitat – cannot
continue to exist.

The effects of stopping the decomposition activities of an
area have been demonstrated, especially in agricultural systems subject to excessive use of pesticides: decomposition,
nutrient release and consequently plant growth comes to a
stop (*e.g.*, Lofty, 1974).

It is hardly necessary to point out that the plants in
turn, especially in forests, have a major influence on both
microclimate and soil properties.Therefore, the activity of
the soil biota can be thought of as a "catalytic" one.
Little energy is converted by the organisms, but the whole
system would change fundamentally in its absence.

Further catalytic effects operate among the soil organisms
themselves. If we consider the different groups of organisms
simply in terms of numbers or of the energy liberation for
which they are directly responsible, we find that bacteria,
for instance, greatly outnumber all other organisms — the
order of 10^{10}–10^{12} per gram of soil. The bacteria and
fungi together are responsible for at least 90 percent of
total energy and carbon dioxide liberation. However, it is
now clear as already mentioned, that most bacteria for most
of the time are inactive presumable because, being immobile
and having highly specialized food demands, they are out of
range of suitable food sources.

It is largely as a result of the activity of the soil
fauna that bacteria are moved to new regions and get the
chance to feed and multiply. Fungi do not have locomotory
organs but they can grow outward from one food source and
colonize another, as anyone familiar with the dry rot fungus
(*Meruluis lachrymans*) can testify. In temperate soils near
neutrality, the metabolic activity of the fungi exceeds that
of the bacteria by two or three times (Gray, 1976). Fungi

can also live in much more acid soils than bacteria, but they in turn are dependent to a large extent on transport of spores by soil animals. It is clear, therefore, that the sum of the action of the soil organisms is greater than that of the component groups of organisms. Again, one must take into account the interactions between these different groups and also their interactions with plant roots if one wishes to assess their contribution to the soil and the ecosystem of which it is part.

REFERENCES

Coleman, D. C. "Soil Carbon Balance in a Successional Grassland," *Oikos* 24:195-199 (1973).

Coleman, D. C. "A Review of Root Production Processes and Their Influence on Soil Biota in Terrestrial Ecosystems," in *The Role of Terrestrial and Aquatic Organisms in Decomposition Processes*, J. M. Anderson and A. Macfadyen, eds. (Oxford: Blackwell, 1976), pp. 417-434.

Gray, T. R. G. "Survival of Vegetative Microbes in Soil," in *The Survival of Vegetative Microbes*, T. R. G. Gray and J. R. Postgate, eds. (Cambridge: Cambridge University Press, 1976), pp. 327-364.

Harley, J. L. "Symbiosis in the Ecosystem," *J. Nat. Sci. Council of Sri Lanka* 1:31-48 (1973).

Jenkinson, D. S. "Studies on the Decomposition of ^{14}C-Labelled Organic Matter in Soil," *Soil Sci.* 111:64-70 (1971).

Macfadyen, A. "Thermal Energy as a Factor in the Biology of Soils," in *Thermobiology*, A. H. Rose, ed. (London and New York: Academic Press, 1967), pp. 535-554.

Macfadyen, A. "Soil Metabolism in Relation to Ecosystem Energy Flow and to Primary and Secondary Production," in *Methods of Study in Soil Ecology*, J. Phillipson, ed. (Paris: UNESCO Symposium, 1970), pp. 167-172.

Lofty, J. R. In *Biology of Plant Litter Decomposition*, C. H. Dickinson and G. H. F. Pugh, eds. (London: Academic Press, 1974), pp. 467-487.

Stout, J. D., K. R. Tate and L. F. Molloy. "Decomposition Processes in New Zealand Soils with Particular Respect to Rates and Pathways of Plant Degradation," in *The Role of Terrestrial and Aquatic Organisms in Decomposition Processes*, J. M. Anderson and A. Macfadyen, eds. (Oxford: Blackwell, 17th British Ecological Society Symposium, 1976), pp. 97-144.

Swaby, R. J. and B. I. Passey. "A Simple Macrorespirometer for Studies in Soil Microbiology," *Aust. J. Agric. Res.* 4:334-339.

CHAPTER 33

CONTRIBUTIONS OF SOIL ORGANIC MATTER IN THE SYSTEM SOIL-PLANT

W. FLAIG

Institut für Biochemie des Bodens
Bundesforschungsanstalt für Landwirtschaft
Bundesallee 50
D-3300 Braunschweig, Federal Republic of Germany

Soil is the upper layer of the earth's crust in which life exists, and it is the substrate on which plants grow. Plants are the basis for food and feedstuff. Therefore, the study of improving the conditions for plant production is very important to mankind, especially in the future, and includes many problems of environmental biogeochemistry.

From a chemical point of view soil consists of inorganic and organic components. Interactions between them both influence plant growth. Moreover, the organisms living in and on the soil have influence on these interactions. The content of organic materials in soils varies between 0.5 and 5 percent. When one bears in mind that the organic soil constituents originate mostly from dead organisms, which consist of a large number of compounds, and that these are transformed in the soil at different rates, it is easy to understand that the number and the concentration of compounds that can be isolated from soil vary greatly according to the conditions. Fractions of humic substances are a relatively stable type of organic soil constituents.

While considering the contributions of soil organic matter in the soil-plant system, the different possibilities of its influence on growth and yield of plants should be noted. We differentiate between *indirect* and *direct* effects. The most important of these are enumerated in Table I. The *indirect* effects change the conditions in the root area. The *direct* effects are those that occur after uptake of constituents of soil organic matter by the roots of plants. Not all items in Table I will be discussed; they are summarized in other publications (Flaig, 1968, 1975).

Table I
Influence on Growth and Yield
of Plants by Constituents of Soil Organic Matter

Indirect Effects (root area)	Direct Effects (after uptake by the roots)
Alterations of *physical* properties (*e.g.*, soil fabric, exchange capacity) Participation in *chemical* reactions (*e.g.*, weathering, complex formation with heavy metals, interactions with pesticides) Alterations of enzyme *activities* in soil	Increase of *uptake* and *distribution* of *heavy metals* in form of complexes *Participation* in *metabolism* in synthesis of plant constituents Influence on *metabolic* pathways (carbohydrate-, protein metabolism, citric acid cycle, ion transport *Release* of N, P and S from *organic linkage*

Because different sources of phenols exist in soil, the possible contributions of this type of compound are mentioned. In addition a brief remark about the release of nutrients from organic linkage has also been made. For instance, studies with [15]nitrogen demonstrated that about 50 percent of the nitrogen in the harvested plant originated from the organic nitrogen compounds of soil organic matter and the other part from the added mineral fertilizer. The nitrogenous organic soil constituents serve as a slow release nitrogen source for plant nutrition, which is considered a favorable factor for plant production.

INFLUENCE OF SOIL ORGANIC MATTER CONSTITUENTS
ON SOME PHYSICAL PROPERTIES OF SOIL

To understand the physical effects of constituents of soil organic matter, the transformations of high molecular weight substances from dead organisms in soil as well as the synthesis of new types of high molecular weight substances formed in soil are summarized in Figure 1. Cellulose, which is the main high molecular weight constituent of plants, is used in the metabolism of soil microorganisms and released to a large part as a metabolite in the form of carbon dioxide. Plant pectins are more or less easily decomposed by the activity of microorganisms. Chitin is an important constituent of some microorganisms for these considerations. Furthermore, some microorganisms synthesize different polysaccharides according to their species.

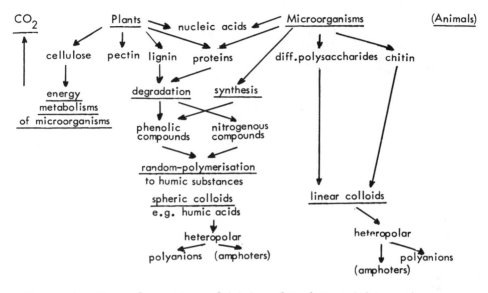

Figure 1. Transformation of high molecular weight, main constituents of organisms in soil organic matter.

Phenolic compounds, which are important building blocks for the synthesis of humic substances, are formed in soil by the degradation of lignin or synthesized by microorganisms from aliphatic precursors. The different fractions of humic substances are formed by random polymerization of phenolic compounds with nitrogenous degradation products of proteins. The humic acids are spheric colloids with the properties of heteropolar polyanions and, to a lesser extent, also amphoters.

In contrast to humic acids, the microbially synthesized polysaccharides are linear colloids. Most of them are heteropolar polyanions. Some may have the properties of amphoters, when for instance glucosamine is a constituent of the high molecular chain. The linear polymers are more effective for interactions with inorganic soil colloids, such as clay minerals, than spheric polymers, such as humic acids (Flaig, 1976).

As a result of different physicochemical investigations, it can be demonstrated in a model experiment with the electron microscope that the mechanisms of the interactions are different (Figure 2). In the case of electrodialized kaolinite, the more or less spheric polymers of the humic acids are not sorbed at the surface of the clay mineral. Interactions occur when the surface of the clay mineral is covered with sesquioxide hydrates or polyvalent cations. In the case of interactions of filamentous and branched high molecular weight linear polyanions with clay minerals, it is not necessary to cover the surface with sesquioxide hydrates or polyvalent cations.

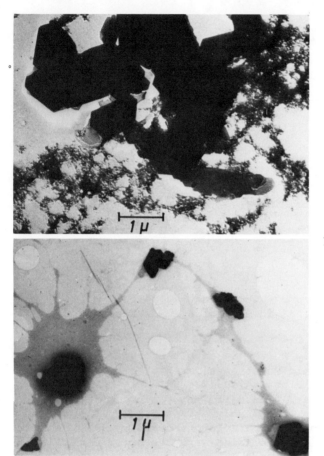

Figure 2. Interactions between kaolinite and humic acids or krilium.

These interactions have a large influence on pore size and distribution and, therefore, on gas exchange, water management, exchange capacity, displacement of soil constituents and finally on plant growth and yield. Because of this it is of interest to study the stability of these two types of polymers in soil. Several experiments have been carried out with [14]carbon-labelled compounds. The rate of decomposition of microbial synthesized polysaccharides in soil is comparable to that of corn stalks used as rural residues for stabilizing soil organic matter in arable soils, while the stability of humic acids was more than ten times higher (Martin *et al.*, 1974, 1975).

DYNAMIC OF FORMATION OF SOIL ORGANIC MATTER

The amount of soil organic matter depends mainly on climatic conditions, mineral parent material and plant cover. Kononova (1966) describes the dependence of humus reserves

in different soils to a depth of 100-120 cm under different
climatic conditions from the north to the south of USSR
(Figure 3). With increasing temperature (a) and decreasing

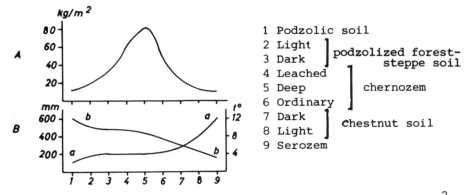

A   The humus reserve to a depth of 100 - 120 cm in kg per m^2.

B   a. Mean annual temperature; b. Mean annual precipitation.

Figure 3.   The humus reserves and climatic conditions of the
   main soil types and subtypes of the USSR.

rainfall (b) a maximum of soil organic matter occurs in the
case of deep chernozems, which decreases again to serozems.
The seasonal dynamics of soil organic matter has an influence
on nutrient cycles in the soil and therefore on the yield
formation of plants. This concerns the naturally occurring
organic materials in soil as well as those added by man in
the form of different rural and urban residual products.
       For elucidation of chemical structure of the system
of organic substances in soil and their dynamics, a chemical
analytical system was developed by us (Salfeld and Sochtig,
1975, 1976). Manifold parameters are necessary for characterization of such a complex mixture of substances. The analytical
data are the basis for quantification of connections between
humus dynamics and agronomical data. Automation of analytical
procedures and electronic data registration are necessary to
obtain enough data for a statistical evaluation.
       An example of the changes of carbon and nitrogen
content of an arable soil is illustrated in Figure 4. Different amounts of straw or green manure were added to the soil
with additional nitrogen fertilization. As a final result,
the polynomial regression curves of the data determined between April and October show that the ratio between nitrogen
and carbon continuously changed in a different way during

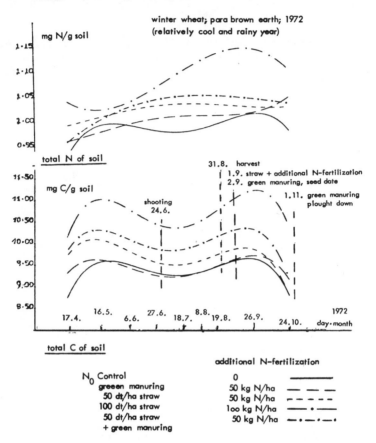

Figure 4.   Carbon and nitrogen of a soil in the course of vegetation.

the dynamic processes in the system of soil organic matter in the course of vegetation period (Sadat, 1974).

PHENOLIC COMPOUNDS AS BUILDING BLOCKS OF HUMIC SUBSTANCES

For more chemical considerations, the formation of humic substances is explained by the scheme given in Figure 5 (Flaig, Beutelspacher and Rietz, 1975; Schnitzer and Kahn, 1972). As mentioned above, the main organic constituents of the dead plants, as the main source for organic materials in soil, are cellulose, lignin and proteins. Different degradation products are formed by microbial activity. Lignin is degraded to phenolic compounds, which are also formed by microbial syntheses (Haider, 1976). The phenolic substances are mainly transformed by oxidation, hydroxylation and decarboxylation. During these transformations ring cleavage of the phenolic compounds also occurs and aliphatic keto acids are formed, which are used by the microorganisms as carbon sources.

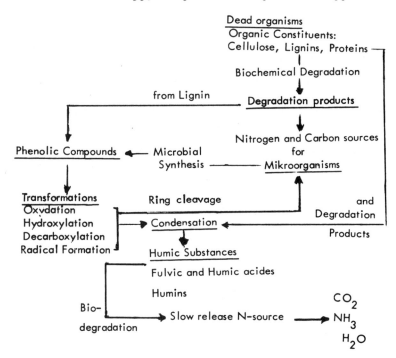

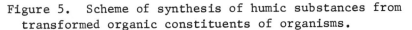

Figure 5. Scheme of synthesis of humic substances from transformed organic constituents of organisms.

During the transformations of the phenolic substances radical formation also occurs. These radicals condense with degradation products of proteins from plants and microorganisms to form humic substances. The humic substances are separated by conventional methods into fulvic and humic acids and humins. The humic substances are not stable; they are biodegraded and serve as slow-release nitrogen sources for plant nutrition. The final degradation products are CO_2, NH_3 and H_2O.

The rate of decomposition of the different plant constituents varies. Cellulose is degraded faster than lignin. The details of chemical reactions during the decomposition of lignins cannot be discussed. These processes were studied with different [14]carbon-labelled synthetic lignins (Martin and Haider, 1976). Only some reactions of lignin degradation products will be mentioned because (a) they can be isolated from humified plant materials (Maeder, 1960; Flaig, 1962), from soil (Bruckert, Jaquin and Metche, 1967; Schnitzer, 1972) and from peat (Belav, 1967; Sochtig and Maciak, 1971), and (b) they have an effect on pathways of plant metabolism (Flaig, 1968).

The most important transformations of lignin degradation products, as shown in Figure 6, are:

1. the shortening of the side chain of phenol acrylic acids and formation of phenol carboxylic acids
2. the demethylation, such as transformation of vanillic acid to protocatechuic acid
3. hydroxylation, such as transformation of hydroxy benzoic acid to protocatechuic acid
4. oxidative decarboxylation, such as transformation of vanillic acid to methoxy-hydroquinone

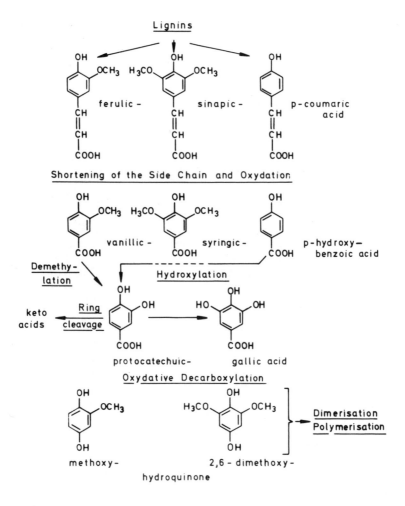

Figure 6.   The main reactions of oxidative transformation of lignin degradation products.

5. the cleavage of the ring. In this way protocatechuic acid is transformed into different keto-acids, which, as mentioned before, are used as a carbon source by the microorganisms.

A second way of synthesizing phenolic compounds during formation of humic substances goes through the metabolism of microorganisms as mentioned above. Most phenols formed in this way belong to 1,3-di- or 1,3,5-triphenols and cannot be derived from lignin or its degradation products. The participation of phenols of the resorcinol type in the formation of humic substances has been demonstrated (Martin and Haider, 1971). For instance *Epicoccum nigrum* forms dark-colored substances with properties comparable to those of humic acids (Filip *et al.*, 1974) (Figure 7).

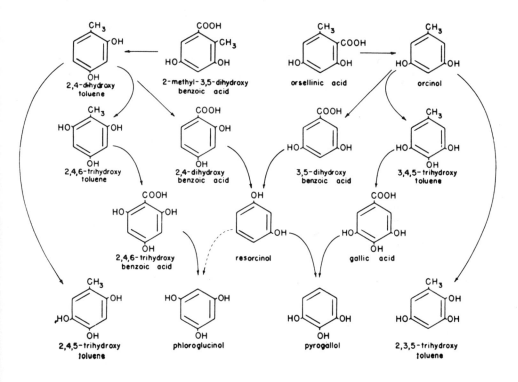

Figure 7. Transformation of microbially synthesized phenols in the culture media of *Epicoccum nigrum*.

Orsellinic and 2-methyl-2,5-dihydroxy-benzoic acid are formed in the metabolism of the fungus from aliphatic carbon sources, *e.g.*, glucose. They are transformed by the main reactions of decarboxylation, oxidation of methyl group to carboxyl group and again decarboxylation, and hydroxylation.

These reactions form several phenolic compounds, with at least two hydroxy groups in the o-position. Such derivatives of polyphenols can be oxidized to quinones in the culture medium at pH values of 6 to 8. These intermediates react with other phenols or nitrogenous substances derived from proteins. This means that generally compounds that are able to form quinones would be important for the formation of nitrogenous humic substances. Other soil fungi also synthesize humic acid-like substances (Haider et al., 1975).

It has been mentioned before that phenolic compounds with hydroxyl groups in the o- or p-position are formed by lignin degradation, while the phenolic compounds formed by microbial synthesis have hydroxyl groups mainly in m-position. These two types of phenolic compounds differ in their reactivity. The main reactions are mentioned in Figure 8. After

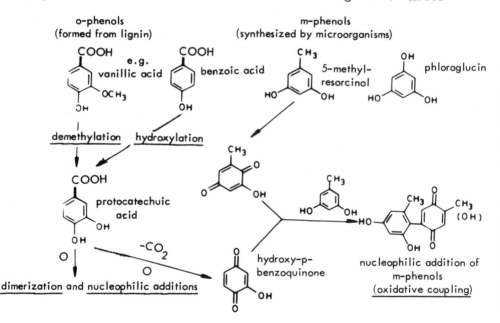

Figure 8. Transformation of polyphenols under oxidizing conditions.

demethylation of lignin or after hydroxylation of p-hydroxy benzoic acid, degradation products such as vanillic acid and some phenolic compounds are formed with several hydroxyl groups in o- or p-position. In the oxidizing medium dimerization and polymerization occur. By oxidative decarboxylation, quinonoid compounds can be formed. During oxidation nucleophilic addition reactions occur through semiquinonoid intermediates.

Many of the microbially synthesized phenols have two or three hydroxyl groups in the m-position. No nucleophilic addition occurs directly either with phenolics or with

nitrogenous compounds. Only during oxidation or after hydroxylation are reactive phenols formed, which add nucleophilicaly
derivatives of resorcinol and phloroglucinol (Musso *et al.*,
1965) through quinonoid intermediates. The oxidative coupling
between phenols leads to ramification and increases the
aromaticity of the humic substances. The differences in
reactivity of the phenolic compounds are also important with
respect to the function of nitrogen in the molecule of humic
substances.

THE NITROGEN CONTENT IN HUMIC ACIDS

In Figure 9 only some reactions are summarized to
elucidate the linkage of nitrogen in humic acids. Guaiacol
and resorcinol derivatives do not add amino acids in a pH-
range of 6.5 to 8.0 in the presence of phenoloxidases in
oxidizing medium. Guaiacol derivatives react under these
conditions to nitrogen-free polymers (Flaig and Haider, 1961).

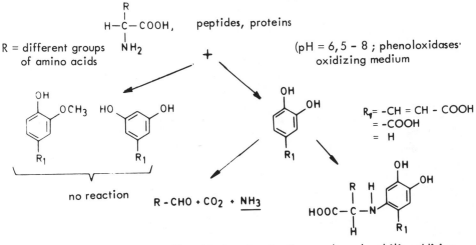

Figure 9. Nucleophilic addition (a) of proteins and degradation products by oxidized phenols; oxidative deamination
(b) of amino acids.

Extensive studies have been made with labelled amino
acids and peptides about the possible linkage of protein-
derived substances with hydroxyl groups in *o*-position by
nucleophilic addition in thecase of phenolic compounds. The
hydrolysis of addition products with proteins or peptides
by 6 *N* hydrochloric acid has shown that all amino acids could

be hydrolyzed with the exception of the N-terminal amino acid,
in which the amino group has reacted with the oxidized phenol
(Haider, Frederick and Flaig, 1965; Haider and Martin, 1970).
Therefore, not all nonhydrolyzable nitrogen is found in humic
substances in heterocyclic form as is often supposed. This
type of reaction is important in connection with the availability of organically bound nitrogen for microorganisms and
plants.

The rate of nucleophilic addition or oxidative deamination depends upon the substitution of the *o*-diphenols. Both
reactions generally occur simultaneously. By oxidative deamination the amino acids are transformed to ammonia, carbon
dioxide and a carbonyl compound (Flaig and Riemer, 1971).
The liberation of ammonia in this way may be interesting in
the processes of humification in relation to plant nutrition.
These studies are contributions to the problem "soil organic
matter as slow release fertilizer" (Flaig and Sochtig, 1967).

UPTAKE OF PHENOLIC COMPOUNDS AS A PREREQUISITE
OF DIRECT EFFECTS OF CONSTITUENTS OF SOIL ORGANIC MATTER

To study quantitatively the uptake, the transport and
the transformation of phenolic lignin degradation products,
such as phenol carboxylic acids in wheat seedling, the
compounds have been labelled at different carbon atoms such
as at the carboxyl group, at the carbon atom of the methoxyl
group or uniformly in the benzene ring according to the
problem to be investigated. The experiments were carried out
in a special apparatus. The plants were prepared by lyophyli-
zation for extraction with different solvents. The activities
of the extracted materials were determined. In this way,
statements could be made about the distribution of the activity
in different plant constituents and the course of reactions
taking place. As one example for such experiments only the
results of the addition of ^{14}C-carboxyl-labelled vanillic
acid are mentioned (Figure 10) (Harms, 1968; Harms *et al.*,
1969 a,b, 1971).

The main results are the following:

1. A relatively high amount of labelled carbon dioxide
   was found, which came from the decarboxylation of
   carboxyl-labelled vanillic acid.
2. The quantity of free vanillic acid in the plant is
   relatively low as indicated by the activity of the
   residues in the ether extracts.
3. The largest part of activity was in the methanol
   extracts. By further investigations it was found that
   vanillic acid was transformed into glucose, into its
   glucose ester or into glycosides of glucose ester.

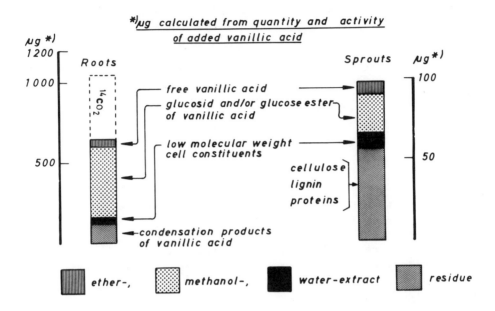

Figure 10. Content of activity in the extracts of roots and sprouts after incubation with carboxyl-labelled vanillic acid (6 days).

4. Some ^{14}C-labelled low molecular weight plant constituents could be identified in the water extracts.
5. In the residues of water extracts of the sprouts, cellulose, lignin and proteins contained radioactivity. This results and that of number 4 can be explained by endogenous fixation of labelled carbon dioxide, which was formed by oxidative decarboxylation of the added ^{14}C-vanillic acid.
6. The activity in the residues of water extracts of the roots was caused by oxidative polymerization of vanillic acid and also with some degradation products.
7. The relatively high release of labelled carbon dioxide is a measure of the oxidative decarboxylation of vanillic acid. By this reaction methoxy-hydroquinone is formed and could be identified in a quantity, which corresponds to about 1 percent of the added activity. For this investigation the vanillic acid used was labelled in the carbon atom of the methyl group or in the ring.

These results are interesting because hydroquinones are 10 to 100 times more active physiologically in plant metabolism, than the corresponding phenol carboxylic acids.

Experiments with plant seedling need a great deal of preparation time. Therefore, we continued these investigations with plant cell suspension cultures. One of the advantages is that the compounds enter the cells by diffusion without all the problems connected with the uptake by roots. Furthermore, it was easier to study the kinetics of the reactions (Berlin *et al.*, 1971; Harms *et al.*, 1972; Harms, 1976).

CONCLUSIONS

The knowledge of the chemical reactions forms an important basis for the elucidation of the causal connections of contributions of soil organic matter in the Soil – Plant system. The results can also be useful for studies in bio-geochemistry of other naturally occurring organic materials.

REFERENCES

Belav, L. *Chemische Untersuchungen heimischer Torfbildner zur Kenntnis der Umwandlung von Pflanzenstoffen unter natürlichen und künstlichen Bedingungen*, Dissertation Rostock (1967).

Berlin, J., K. Barz, H. Harms and K. Haider. "Degradation of Phenolic Compounds in Plant Cell Cultures," *FEBS Letters* 16:141-146 (1971).

Bruckert, S., F. Jaquin and M. Metche. "Contribution à l'étude des acides phenoles présents dans les sols," *Bull. d'école Nationale Supérieure Agronomique de Nancy* IX(II): 73-92 (1967).

Filip, Z., K. Haider, H. Beutelspacher and J. P. Martin. "Comparisons of RR-Spectra from Melanins of Microscopic Soil Fungi, Humic Acids and Model Phenol Polymers," *Geoderma* 11:37-52 (1974).

Flaig, W. "Zur Umwandlung von Lignin in Humusstoffe," *Freiberger Forschungshefte A* 254:39-56 (1962).

Flaig, W. "Uptake of Organic Substances from Soil Organic Matter by Plant and Their Influence on Metabolism," *Pontificiae Academiae Scientiarum Scripta Varia* 32:723-770 (1968).

Flaig, W. "An Introductory Review on Humic Substances: Aspects of Research on Their Genesis, Their Physical and Chemical Properties, and Their Effect on Organisms," *Proc. Int. Meet. Humic Substances, Nieuwersluis* (1975).

Flaig, W. "Contributions of Biochemistry of Soil Organic Matter to Soil Conditioning, Med. Fac. Landbouw. Rijksuniv. *Gent* 41:23-40. (1976).

Flaig, W. and K. Haider. "Reaktionen mit oxydierenden Enzymen aus Mikroorganismen," *Planta Medica, Z. f. Arzneipflanzenforsch* 9:123-139 (1961).

Flaig, W. and H. Söchtig. "Organische Verbindungen als Stickstoffquelle fur die Ernährung der Pflanzen," *Anales de Edafologia y Agrobiologia* XXVI:801-828 (1967).

Flaig, W. and H. Riemer. "Polarographische Untersuchungen zum Verhalten von Trihydroxytoluolen bei der Reaktion mit Glycin unter oxydierenden Bedingungen," *Justus Liebig Annalen der Chemie* 746:81-85 (1971).

Haider, K. "Microbial Synthesis of Humic Materials," *American Chemical Society (ACS) Series* 29:244-257 (1976).

Haider K., L. R. Frederick and W. Flaig. "Reactions Between Amino Acid Compounds and Phenols During Oxidation," *Plant and Soil* XXII:49-64 (1965).

Haider, K. and J. P. Martin. "Humic Acid-Type Phenolic Polymers from Aspergillus Sydowi Culture Medium. Stachybotry spp. Cells and Autoxydized Phenol Mixtures," *Soil Biol. Biochem.* 2:145-156 (1970).

Haider, K., J. P. Martin, Z. Filip and E. Fustec-Mathon. "Contribution of Soil Microbes to the Formation of Humic Compounds," *Proc. Int. Meet. Humic Substances, Nieuwersluis,* Center for Agricultural Publishing and Documentation, Wageningen, Netherlands, pp. 71-85 (1975).

Harms, H. "Über die quantitative Bestimmung der Aufnahme von Phenolcarbonsäuren durch die Wurzeln von Weizenkeim-pflanzen," *Mitt. Dtsch. Bodenkundl. Ges.* 8:159-161 (1968).

Harms, H. "Metabolism of Soil Related Phenolic Compounds in Plants and Cell Suspension Cultures," *Proc. Int. Symp. "Soil Organic Matter Studies" IAEA, Vienna* 1976 Volume 2 (in press).

Harms, H., H. Söchtig and K. Haider. "Untersuchungen zur Aufnahme und Umwandlung C^{14}-markierter Phenole durch die Pflanze. I. Aufnahme von C^{14}-carboxylmarkierter p-Hydroxybenzoe-, Vanillin- und Syringasaure durch die Wurzeln von Weizenkeimpflanzen und Verteilung der Aktivitat uber die Pflanze," *Plant and Soil* XXXI:129-142 (1969).

Harms, H., H. Söchtig and K. Haider. "Untersuchungen zur Aufnahme und Umwandlung C^{14}-markierter Phenole durch die Pflanze. II. Die Umwandlung von p-Hydroxygenzoesaure, Vanillinsäure sowie Syringasaure nach der Aufnahme durch die Wurzeln von Weizenkeimpflanzen," *Plant and Soil* XXXI:257-272 (1969).

Harms, H., H. Söchtig and K. Haider. "Aufnahme und Umwandlung von in unterschiedlichen Stellungen C^{14}-markierten Phenolcarbonsäuren in Weizenkeimpflanzen," *A. Pflanzenphysiol.* 64:437-445 (1971).

Harms, H., K. Haider, J. Berlin, P. Kiss and W. Barz. "Uber O-Demethylierung und Decarboxylierung von Benzoesäuren in pflanzlichen Zellsuspensionskulturen," *Planta* 105:342-351 (1972).

Kononova, M. M. *Soil Organic Matter, Its Nature, Its Role in Soil Formation and in Soil Fertility*, 2nd Edition, Pergamon Press, Oxford, London, Edingburgh, New York Toronto, Sydney, Paris, Braunschweig) (1966).

Maeder, H. *Chemische und pflanzenphysiologische Untersuchungen mit Rottestroh* Diss. Justus Liebig Universitat Gie en (1960).

Martin, J. P. and K. Haider. "Microbial Activity in Relation to Soil Humus Formation," *Soil Sci.* 111:54-63 (1971).

Martin, J. P., K. Haider, W. J. Farmer and E. Fustec-Mathon. "Decomposition and Distribution of Residual Activity of Some ^{14}C-Microbial Polysaccharides and Cells, Glucose, Cellulose and Wheat Straw in Soil," *Soil Biol. Biochem.* 6:221-230 (1974).

Martin, J. P., K. Haider and E. Bondietti. "Properties of Model Humic Acids Synthesized by Phenooxidase and Autoxidation of Phenols and Other Compounds Formed by Soil Fungi," *Proc. Int. Meet. Humic Substances, Nieuwersluis* pp. 171-186 (1975).

Martin, J. P. and K. Haider. "Decomposition in Soil of Specifically ^{14}C-Labelled DHP and Corn Stalk Lignins, Model Humic Acid-Type Polymers and Coniferyl Alcohols,"

Proc. Int. Symp. "Soil Organic Matter Studies" IAEA, Vienna, Volume 2 (in press).

Musso, H., U. V. Gizycki, H. Kramer and H. Dopp. "Über Orceinfarbstoffe. 24. Über den Autoxydationsmechanismus bei Resorcinderivaten," *Chem. Ber.* 98:3952-3963 (1965).

Sadat-Dastegheibi, B. *Untersuchungen zur Stoffdynamik in Ackerböden in Abhängigkeit von Verschiedener Organischer Düngung und Stickstoffdüngung* Dissertation Justus-Liebig-Universität, Fesse Giessen (1974).

Salfeld, J.-Chr. and H. Söchtig. "Proposals for the Characterization of Soil Organic Matter as an Approach to Understand Its Dynamics," *FAO Soils Bulletin* 27:71-81 (1975).

Salfeld, J.-Chr. and H. Söchtig. "Composition of the Soil Organic Matter System Depending on Soil Type and Land Use," *Proc. Int. Symp. "Soil Organic Matter Studies" IAEA, Vienna,* Volume 1:227-235 (1976).

Schnitzer, M. and S. U. Khan. *Humic Substances in the Environment* (New York: Marcel Dekker, Inc., 1972) pp. 327.

Söchtig, H. and F. Maciak. "Bindung des Stickstoffs und Vorkommen Phenolischer Verbindungen im Torf," *Telma* 1:49-61.

SECTION II

PADDY SOILS, PEAT AND COAL

CHAPTER 34

MICROBIOLOGY OF PEAT

E. KÜSTER

Department of Agricultural Microbiology
University Giessen
Giessen, Federal Republic of Germany

Peat is an accumulation of organic matter, resulting from the transformation of the original vegetation under partially anaerobic conditions. This plant material gradually underwent an incomplete decomposition. The kind of peat formed depends on the original plant association, the climate at the time of formation, and the microorganisms active in the decomposition process. Consequently, the chemical composition varies from one type of peat to another. In the course of carbonization of organic matter, peat is a stage between humic acids on the one hand and lignite, coal and anthracite on the other. There is also a differentiation made between low-moor and high-moor peat. The first is formed on eutrophic waters and, consequently, is rich in ash and nutrients. The high-moor peat is the product of partially decomposed mosses (sphagnum), grasses (sedges, eriophorum) and the like, and is distinguished by its extremely low ash and nutrient content.

The microbiology of peat is a wide field, which can be considered under various aspects.

1. The occurrence and activity of microorganisms present in virgin bogs in their original state.
2. The composition of the microflora and their metabolism in harvested and more or less processed peat.
3. The influence of microorganisms on peat.
4. The effect of peat and its components on microorganisms and microbial activities.

The first two points are ecological ones; they cannot be separated from the following two, which emphasize the mutual relationship between substrate and organisms. A short review of the first three points follows.

The old conception that peat is more or less a sterile material without considerable microbial life and activity has been revised in the last 40–50 years. The number of micro-organisms found in peat is much higher than originally assumed. There is also some evidence that microorganisms in peat are not only in a dormant state, but are also actively living. Many microbial processes of transformation and decomposition of peat substances take place in the natural habitat. This means that to a certain extent peat substances must be used as nutrients and sources of energy. The more decomposed, humified and carbonized the carbon material in a peat bog, the smaller the content of utilizable compounds and the less available they become for microbes (Waksman and Purvis, 1932). The availability of carbon substances for microbial develop-ment decreases with increasing carbonization in the order humic acids, peat, lignite, coal, anthracite (de Barjac, 1953). Considerable amounts of hemicelluloses and cellulose have been found in high-moor peat, the less decomposed layers con-taining more of these substances than the well-decomposed ones (Waksman and Stevens, 1928).

According to the high acidity of peat, fungi are in the majority, but also present are actinomycetes and bacteria that are adapted to this extreme habitat. The discrepancy in the studies on the composition of the fungal as well as of the bacterial flora of peat samples is probably due to both the different techniques used and a real difference between the microflora of each sample. Peat is a very hetero-genous material, varying by the kind of its formation and by its botanical origin. According to the differences in the chemical composition and the physical properties, the micro-flora also vary. As opposed to samples of mineral soils, peat samples contain only a few different varieties of organ-isms. *Penicilli* were the predominant fungi in the peat samples. An addition, *Aspergillus* spp., *Cephalosporium, Pullularia, Trichoderma* and a number of other genera have been isolated (Boswell and Sheldon, 1951; Dahmen and Ramaut, 1954; Moore, 1954; Kendrick, 1962; Christensen and Whittingham, 1965; Todorovic and Bogdanovic, 1967; Zabawski, 1967; Dickinson and Dooley, 1969; Dooley, 1970). Phycomycetes are generally rare; *Mortierella* and *Pythium* replace the other-wise common genera *Mucor* and *Rhizopus* as phycomycete genera (Robertson, 1973). The frequent occurrence of *Cladosporium, Aleurisma, Cephalosporium, Trichoderma*, and others is probably due to their strong cellulolytic activity (Boswell and Sheldon, 1951; Lobanok and Shklyar, 1968). *Cladosporium, Penicillium* and others are also considered to be mycorrhizal fungi on typical bog plants such as Calluna, Empetrum and Vaccinium. It has also been observed that sphagnum is only able to grow on sterile peat in the presence of a *Penicillium* species (Burgeff, 1956).

With regard to the small number of varieties, the same also occurs within the bacterial flora (Beck and Poschenrieder, 1961). Spore-forming *Bacillus* sp. are reported to be in the majority (60-80 percent) (Holding *et al.*, 1965; Latter *et al.*, 1967). In other samples Gram-negative rods, particularly *Pseudomonas*, are predominant (Beck and Poschenrieder, 1958; Poschenrieder, 1958; Poschenrieder and Beck, 1958; Roschenthaler and Poschenrieder, 1958; Janota-Bassalik, 1963; Christensen and Cook, 1970). In addition, strains of the genera *Micrococcus, Arthrobacter, Achromobacter, Brevibacterium, Chromobacterium, Mycobacterium*, and coryneforms have been found. New microbial forms were recently registered in peat soils by electron microscopy. These forms were characterized as star-like bacteria with appendages and flat microorganisms with pores. Some of those are considered to be used for diagnosis of the peat type (Volarovich and Terentjev, 1970). Poschenrieder and Beck (1958) suggest that the *Pseudomonades* exhibit a particular importance for the peat formation. The composition of the bacterial flora of peat is very similar to that of lignite and brown coal (Beck *et al.*, 1955; Guthenberg, 1958; Jaschhof and Schwartz, 1969). In a bog under extremely antarctic conditions Brevibacteria were recorded as predominant (>50 percent); also present were *Arthrobacter, Cellulomonas, Kurthia* and *Micrococcus*. It is surprising that the number of facultative psychrophiles in that habitat does not exceed 60-65 percent (Baker and Smith, 1972). Comparing two high-moor samples from different areas, Upper Bavaria and Lower Saxonie, it was stated that the microflora of these bogs are nearly equal in their qualitative composition (Rehm and Sommer, 1962).

Due to the anaerobic conditions which are predominant in the virgin and undrained bogs, the number of strictly aerobic organisms is very low. It has been observed that no nitrification takes place in the untouched bogs of the high-moor type (Kuster and Gardiner, 1968). This is due to the absence of nitrifying bacteria and not to the missing oxidizable ammonia. Herlihy (1972, 1973) recently confirmed this observation in that the obligate autotrophic *Nitrosomonas* is entirely absent, whereas the facultative autotrophic *Nitrobacter* was occasionally present in very low numbers. The anaerobic denitrifiers, on the other hand, are present in bogland. They are not very active in the original environment because the peat is lacking in original nitrates. The ratio of aerobic to anaerobic bacteria varies from 4:1 to 20:1 according to whether the soil was high-moor, transitional or low-moor (Zimenko, 1956).

Peat is similar to other soil profiles in that the number of microorganisms decreases with increasing depth. But the changes in the microflora of peat are somewhat different from those in cultivated fields. Microbial counts of samples from peat profiles showed that the number of bacteria did not decrease steadily, but increased again at 100 cm with a maximum at 150-175 cm. This second maximum coincides with a change in peat structure and grade of decomposition (from H5 to H7-8) (Poschenrieder *et al.*, 1957; Kuster, 1963). This phenomenon is closely related to an increase in anaerobes, absolutely as well as proportionally. The figures of anaerobes in percentage of the total bacterial count gradually increase with depth, from 3-5 percent on the surface to 40-50 percent at 160-185 cm (Waksman and Purvis, 1932; Poschenrieder *et al.*, 1957; Rehm and Sommer, 1962; Rogers, 1966; Stout, 1971). Beck and Poschenrieder (1958) noted that *Pseudomonas* is the only bacterium occurring at a depth of 185 cm and is consequently considered to belong to the autochthonic microflora of peat. It is not surprising that anaerobic bacterial types and their activity were the particular aim of many studies on peat microbiology.

The formation of H_2S is well known in anaerobic zones of the high-moor (Benda, 1957). H_2S derives from the activity of sulfate-reducing bacteria, *e.g.*, *Desulfovibrio*, which are obligate anaerobes. The fact that these organisms, which prefer a neutral reaction, were frequently found in the strongly acidic peat, can be explained by the formation of microhabitats with particularly favorable environmental conditions.

Another fermentation product that occurs to quite an extent in certain bogs is methane. The biological formation of methane is regarded to be associated with the anaerobic cellulose decomposition. In fact, this reaction takes place in peat from which anaerobic cellulose decomposers and methane-forming bacteria have also been isolated (Rogers, 1966).

Actinomycetes are rarely present in virgin bogs because of their acidity and poor aeration. Occasionally, extensively branched and very narrow hyphal forms were observed; they were considered to be vegetative structures of *Nocardiae* (Dickinson *et al.*, 1974).

The occurrence of blue-green algae (*Cyanobacteria*) in peat is very doubtful. A so-called "Moorschnecke" (Leptobasis) has been described in the sphagnetum of a high-moor peat (Hulsbruck, 1967). In so far as blue-green algae have been found in peat, they are mainly present in an inactive and dormant state (Dooley and Houghton, 1973).

The occurrence and frequency of algae in bog water depend on the water's reaction and nutrient status. *Euglenae*, green algae, *Volvocales*, *Desmidiales*, and diatoms have been found (Cholnoky and Schindler, 1953; Loub, 1953; Fetzman, 1963; Fott and McCarthy, 1964).

For the microbiological critical examination of an habitat, it is more important to know the kind and strength of microbial metabolic activities than the number of organisms, or more exactly the number of cells that are able to germinate. For its known chemical and physical properties, peat is an extremely unsuitable substrate for an active microbial life. However, microbial activity in peat should not be neglected although it does not appear to be extensive. Over the extended time scale involved in the accumulation of such peat deposits, the changes that may result from a very reduced level of activity would likely be significant in terms of the chemical and physical properties of this type of sediment (Dickinson *et al.*, 1974).

Pectines play an important role in the structural composition of sphagnum. Together with cellulose and hemicellulose they are peat components that are quickly decomposed and do not accumulate in peat, contrary to lignin and humic substances. Fungi are mainly responsible for that kind of decomposition (Kox, 1954; Frercks and Puffe, 1961). A restricted microflora and the consequent low microbial activity result from a lack of suitable food substrates rather than an adverse physical environment.

After removal of the peat layers for further processing, a last residual peat layer (40-60 cm) is left, overlying the subsoil of various types. Special microbiological problems arise in those cut-away peats, which were studied extensively by Dickinson and Dooley. When a peat area is cut away, the content of species of microbes is greatly reduced (Dickinson and Dooley, 1969). This reduction is a permanent feature of cut-away peat and reflects the lack of higher plant colonization. The microbial activity of cut-away peat is very low when brought into cultivation. Thus, the danger exists that disease organisms could be introduced and become established (Dooley and Dickinson, 1971). Before any intensive crop production is started on peat, the cosmopolitan element of the peat microflora must have been sufficiently encouraged so as to be able to resist invasion by alien species and to support plant growth.

The microbial activity in virgin bogs is originally small and slow. Their drainage, liming, and fertilization change the microflora quantitatively. Nitrification does not occur in uncultivated raised bogs, nitrifiers are completely absent. This is not surprising, because nitrifying organisms present and active in soil indicate a high grade of soil fertility. Nitrification takes place in a well-aerated soil of slightly acidic or neutral reaction. However, these conditions do not occur in peatland. A nitrification is demonstrable only after liming and cultivation for a number of years, but not immediately after addition of calcium and other fertilizers (Kuster and Gardiner, 1968). Herlihy (1972) detected *Nitrosomonas* and *Nitrobacter* in reasonable numbers after only three years of fertilization and cultivation; the number of bacteria and actinomycetes also increased obviously (ten times). While counts of all organisms increased in cultivated peats, the magnitude of change was much greater for the chemoautotrophic nitrifiers, suggesting that the substrate was less limiting for these (Herlihy, 1973). The nitrogen-fixing *Azotobacter* and *Clostridium* are present in cultivated peat although the reaction was only slightly changed (pH 4.5 to 5.0) after cultivation (Remenets and Kudjakova, 1957). Even in rheotrophic peat nitrogenase activity was detected. This could be explained by the existence of nitrogen-fixing associations of aerobic and anaerobic bacteria in peat (Waughman, 1976).

Enzymatic activity is one of the means by which the beneficial effect of the soil microflora can be determined. An acid phosphatase with strong activity has been found in acid high-moor peat (Herlihy, 1972). The activity of various enzymes is changed by liming and fertilization (Kuster and Gardiner, 1968). The microbial respiration in peat, measured as CO_2 produced, is also influenced by the addition of Ca or N. No effect was recorded when both the elements were given in combination (Kuster, 1970).

Heat is evolved by microbial respiration. Heat can accumulate under unfavorable conditions when the aeration system in soil is disturbed. This can happen in a pile of harvested, particularly milled peat. By this kind of processing of peat, *i.e.*, improvement of aeration and other physical factors, the above described biological situation in peat is entirely changed. The number and activity of microorganisms increase.

The best indication of a microbial activity, even for nonmicrobiologists, is the phenomenon of self-heating, which means a rise in temperature inside the material independent of the outside atmospheric termperature. The increase in temperature is the result of microbial respiration caused by a biological process. Plants and microbes respire and develop heat, which is usually drawn off. Under unfavorable physical conditions (compression, moisture, aeration), the heat developed accumulates in small nests. In that microhabitat the temperature continuously increases. With increasing temperatures, the composition of the microflora changes, thermophiles succeed the mesophiles, or more exactly expressed, the ratio of mesophiles to thermophiles is changed in favor of the latter (Kuster, 1969). Beyond a certain temperature (45-50°C) the number of active mesophiles drops and the thermophiles show their full activity. At further increased temperature, higher than 75-80°C, even the thermophiles become inactivated and chemical reactions lead to a self-ignition.

By the heating process and the metabolic activities of the thermophilic microflora in particular, the chemical and biochemical properties of peat are altered. Inhibiting as well as stimulating substances may be produced that exhibit a physiological effect on the growth of plants (Niggemann, 1968). Water extracts of differently heated peat samples were tested against *Azotobacter* (Niese, 1975). The antibiotic activity becomes stronger with increasing temperatures (>75°C), indicating that the temperature level reached and the period of time at which the maximum temperature is kept may influence the formation and effectiveness of peat substances.

In summary, we can state that most of the biochemical properties of peat are induced by microbial activities.

REFERENCES

Baker, J. H. and D. G. Smith. "The Bacteria in an Antarctic Peat," *J. Appl. Bact.* 35:589-596 (1972).

Beck, T., F. Bukatsch and H. Poschenrieder. "Bakterien aus Oberpfälzer Braunkohle und ihr physiologisches Verhalten," *Phyton* 5:79-91 (1955).

Beck, T. and H. Poschenrieder. "Über die artenmäßige Zusammensetzung der Mikroflora eines sehr sauren Waldmoorprofiles," *Ztrlbl. Bakt. II* 111:672-683 (1958).

Beck, T. and H. Poschenrieder. "Wechselwirkungen von Biotop und Bodenmikroflora im Bereich des Hochmoores," *Bayr. Landw. Jahrb.* 38:110-122 (1961).

Benda, I. "Mikrobiologische Untersuchungen über das Auftreten von Schwefelwasserstoff in den anaeroben Zonen des Hochmoores," *Arch. Mikrobiol.* 27:337-374 (1957).

Boswell, J. G. and J. Sheldon. "The Microbiology of Acid Soils, II. The Ringinglow Bog, Near Sheffield," *New Phytol.* 50:172-178 (1951).

Burgeff, H. "Mikrobiologie des Hochmoores, mit besonderer Berücksichtigung der Ericaceen-Pilzsymbiose," *Ber. Dtsch. Bot. Ges.* 69:257-262 (1956).

Cholnoky, B. V. and H. Schindler. "Die Diatomeengesellschaft der Ramsauer Torfmoore," *Sitz. Ber. Österr. Akad. Wiss. Math. Nat. Kl. Abt. I* 162:235-271 (1953).

Christensen, M. and W. F. Whittingham. "The Soil Microfungi of Open Bogs and Conifer Swamps in Wisconsin," *Mycologia* 57:882-896 (1965).

Dahmen, M. and J. Ramaut. "Chemical Composition of Certain Types of Acid Soils and Their Fungus Flora," *Arch. Inst. Bot. Univ. Liege* 22:No. 14 (1954).

De Barjac, H. "Pouvoir inhibiteur des tourbes sur la croissance des germes du sol," *Abstr. VI. Int. Congr. Microbiol. Roma* 3:115 (1953).

Dickinson, C. H. and M. Dooley. "Fungi Associated with Irish Peat Bogs," *Proc. Roy. Irish Acad. Sect.* B68:119-134 (1969).

Dickenson, C. H., B. Wallace and P. H. Given. "Microbial Activity in Florida Everglades Peat," *New Phytol* 73:107-113 (1974).

Dooley, F. and J. A. Boughton. "The Nitrogen-Fixing Capacities and the Occurrence of Blue-Green Algae in Peat Soils," *Brit. Phycol. J.* 8:289-293 (1973).

Dooley, M. "The Microbiology of Cut-Away Peat. IV. Autecological Studies," *Plant and Soil* 33:145-160 (1970).

Dooley, M. and C. H. Dickinson. "The Ecology of Fungi in Peat," *Irish J. Agric. Res.* 10:195-206 (1971).

Fetzmann, E. "Zur Algenflora zweier steirischer Moore," *Protoplasma* 57:334-343 (1963).

Fott, B. and A. J. McCarthy. "Three Acidophilic Volvocine Flagellates in Pure Culture," *J. Protozool.* 11:116-120 (1964).

Frercks, W. and D. Puffe. "Zur Frage des Pektin- und Zelluloseabbaues in Moorböden," *Z. Pflzern. Düngg. Bodenkd.* 92:46-56 (1961).

Guthenberg, H. "Microbiology of Coal," *TVF* 29:216-222 (1958).

Herlihy, M. "Microbial and Enzyme Activity in Peats," *Acta Horticult. Techn. Commun.* No. 26:45-50 (1972).

Herlihy, M. "Distribution of Nitrifying and Heterotrophic Microorganisms in Cutover Peats," *Soil Biol. Biochem.* 5:621-628 (1973).

Holding, A. J., D. A. Franklin and B. Watling. The Micro-flora of Peat-Podzol Transition," *J. Soil Sci* 16:44-59 (1965).

Hülsbruch, M. "Die 'Moorschnecke' (*Leptobasis goesingense*) im Heideweiher des Naturschutzgebietes Heiliges Meer bei Hopsten," *Ber. Dtsch. Bot. Ges.* 80:367-370 (1967).

Janota-Bassalik, L. "Psychrophiles in Low-Moor Peat," *Acta Microbiol. Polon.* 12:25-54 (1963).

Jaschhof, H. and W. Schwartz. "Untersuchungen zur Geomikrobiologie der Braunkohle. I. Die Mikroben-Assoziationen eines Braunkohlenflözes," *Z. Allg. Mikrobiol.* 9:103-119 (1969).

Kendrick, W. B. "Soil Fungi of a Copper Swamp," *Canad. J. Microbiol.* 8:639-647 (1962).

Kox, E. "Der durch Pilze und aerobe Bakterien veranlaßte Pektinabbau und Celluloseabbau im Hochmoor unter besonderer Berücksichtigung des Sphagnumabbaues," *Arch. Mikrobiol.* 20:111-140 (1954).

Küster, E. "Studies on Irish Peat Bogs and Their Microbiology," *Microbiol. Espan.* 16:203-208 (1963).

Küster, E. and J. J. Gardiner. "Influence of Fertilizers on Microbial Activities in Peatland. *III .Int. Peat Congr.* Quebec 2:314-317 (1968).

Küster, E. "Peat Microbiology," *Proc. Biochem.* 4(11):47-50 (1969).

Küster, E. "Der Einfluß von Ca- und N-Gaben auf mikrobielle Aktivitäten in Moorboden und Torf. *Landw. Forschg. Sonderh.* 25/II:115-124 (1970).

Latter, P. M., J. B. Cragg and O. W. Heal. "Comparative Studies on the Microbiology of Four Moorland Soils in the Northern Pennines," *J. Ecol.* 55:445-464 (1967).

Lobanok, A. G. and B. K. Shklyar. "Cellulolytic Activity of Fungi Isolated from Peat," *Ber. Akad. Wiss. BSSR* 3:108-110 (1968).

Loub, W. "Zur Algenflora der Lungauer Moore," *Sitz. Ber. Osterr. Akad. Wiss. Math. Nat. Kl. Abt. I* 162:545-569 (1953).

Moore, J. J. "Some Observations on the Microflora of Two Peat Profiles in the Dublin Mountains," *Proc. Roy. Dublin Soc.* 26:379-395 (1954).

Niese, G. Unpublished results (1975).

Niggemann, J. "Selbsterwärmung von Torf in Mieten und deren Auswirkungen," *Torfnachr.* 19:7-8 (1968).

Poschenrieder, H., O. Kraemer and T. Beck. "Untersuchungen über die mikrobiologischen Verhältnisse in einem sehr sauren Waldmoorprofil," *Mitt. Moor-Torfwirtsch.* 4:28-37 (1957).

Poschenrieder, H. "Die Mikrobiologie der Moore," *Mitt. Moor-Torfwirtsch.* 5:41-51 (1958).

Poschenrieder, H. and T. Beck. "Untersuchungen über die Rolle einiger bei den ersten Stadien des Torfbildungsvorganges beteiligten Bakterienarten," *Ztrlbl. Bakt. II* 111:684-695 (1958).

Rehm, H. J. and G. Sommer. "Mikrobiologische und chemische Untersuchung eines nordwestdeutschen Hochmoores," *Ztrlbl. Bakt. II* 115:594-600 (1962).

Remenets, M. F. and N. D. Kudjakova. "A Study of the Dynamics of Cellulose Decomposition and Fixation of Atmospheric Nitrogen by Microorganisms in Low-Moor Peat," *Ber. Akad. Wiss. Ukrain. SSR* 5:521-524 (1957).

Robertson, G. I. "Occurrence of *Pythium* spp. in New Zealand Soils, Sands, Pumices, and Peat, and on Roots of Container-Grown Plants," *N. Z. J. Agric. Res.* 16:357-365 (1973).

Rogers, J. "Studies on Anaerobes in Peat and on the Effect of Some of Its Extracts on Acetone Production by *Clostridium Acetonbutylicum*," *M. Sc. Thesis Dublin (N.U.I.)* (1966).

Roschenthaler, R. and H. Poschenrieder. "Untersuchungen über die Bakterienflora eines Hochmoorprofiles bei Staltach in Bayern," *Ztrlbl. Bakt. II* 111:653-671 (1958).

Todorovic, M. and V. Bogdanovic. Contribution to the Study of Common Fungi from Some Mountain Peat Bogs," *Mikrobiologija* 4:105-110 (1967).

Volarovich, M. D. and A. A. Terentjev. "New Microbial Forms and Their Distribution in Natural Peat," *Mikrobiologija* 39:488-494 (1970).

Waksman, S. A. and K. R. Stevens. "Contribution to the Chemical Composition of Peat. II. Chemical Composition of Various Peat Profiles," *Soil Sci.* 26:239-251 (1928).

Waksman, S. A. and E. R. Purvis. "The Microbiological Population of Peat," *Soil Sci.* 34:95-109 (1932).

Waughman, G. J. "Investigations of Nitrogenase Activity in Rheotrophic Peat," *Canad. J. Microbiol.* 22:1561-1566 (1976).

Zabawski, J. "Studies on the Mycoflora of Sphagnum Bog "Zieleniec'," *Proc. Polon. Acad. Sci.* 76:355-400 (1967).

Zimenko, T. G. "Distribution of Anaerobic Bacteria in Peat-Bog Soils," *Bull. Inst. Biol.* 2:165-167 (1956).

CHAPTER 35

MICROBIAL ACTIVITY IN DIFFERENT TYPES OF MICROENVIRONMENTS IN PADDY SOILS

Y. DOMMERGUES

Centre de Pédologie Biologique du CNRS
B.P. 5 - 54500 Vandoeuvre-les-Nancy, France

Orstom - B. P. 1386 - Dakar, Senegal

GENERAL CHARACTERISTICS OF PADDY SOILS

Most studies on paddy soils have hitherto been concerned by the overall effect of submergence on biological and chemical soil properties (Redman and Patrick, 1965; Turner and Patrick, 1968). The sequence of oxido-reduction changes in soils that occur as a result of waterlogging together with transformation of manganese from the easily reducible form to the exchangeable form (Mn^{++}), reduction of Fe, release of phosphorus from the non extractable to the extractable form and eventually formation of $S^=$, H_2, CH_4 has been studied by different authors. Figure 1 is a classical example of the effect of waterlogging on some chemical characteristics of a soil.

In an excellent review, Yoshida (1975) clearly emphasized the classical time sequence of the operations usually performed in paddy fields: (1) submergence of the soil, with or without puddling, for the duration of the crop, with or without soil drying in midseason, (2) draining and drying the soils before harvest, and (3) reflooding for the next crop a few weeks to several months after harvest. Takai et al., quoted by Yoshida (1975), studied the successive reducing processes that occurred after the submergence (Table I).

Actually, it has not been sufficiently emphasized that paddy soils, even in the submergence phase, are far from being uniformly at a given reducing level. Paddy soils should be regarded as complex systems formed up by the juxtaposition of microenvironments which are either sites of oxidation reactions or sites of reduction reactions mediated by a host of soil microorganisms. This chapter will present the different categories of environments occurring in paddy soils.

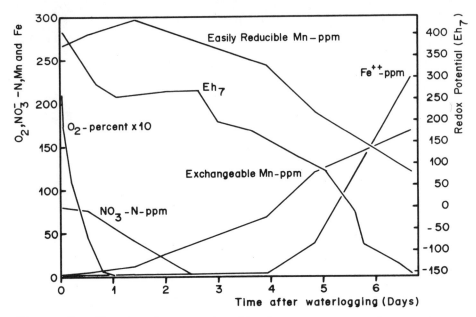

Figure 1.   Changes in oxygen, nitrate, manganese, iron and
   redox potential in a silty clay as a result of waterlogging
   (Turner and Patrick, 1968).

Table I
Successive Microbial Reduction Processes
in Flooded Soils[a]

Transformation of Elements	Initial Soil Eh	Biochemical Pattern
Disappearance of NO^-_3	$+ 0.6 \sim + 0.5$	Aerobic respiration
Disappearance of NO_3	$+ 0.6 \sim + 0.5$	Anaerobic respiration
Formation of Mn^{2+}	$+ 0.6 \sim + 0.4$	Anaerobic respiration
Formation of Fe^{2+}	$+ 0.6 \sim + 0.3$	Anaerobic respiration
Formation of S^{2-}	$0 \sim - 0.19$	Anaerobic respiration
Formation of H_2	$- 0.15 \sim - 0.22$	Fermentation
Formation of CH_4	$- 0.15 \sim - 0.19$	Fermentation

[a]Yoshida, 1975.

MICROBIAL ACTIVITIES IN LARGER MICROENVIRONMENTS

Three layers are usually reported in a submerged paddy
soil (Figure 2). These are the liquid layer (flood water),
the aerobic soil layer and aerobic-anaerobic interface, and
the anaerobic soil layer. Aquatic plants and rice are

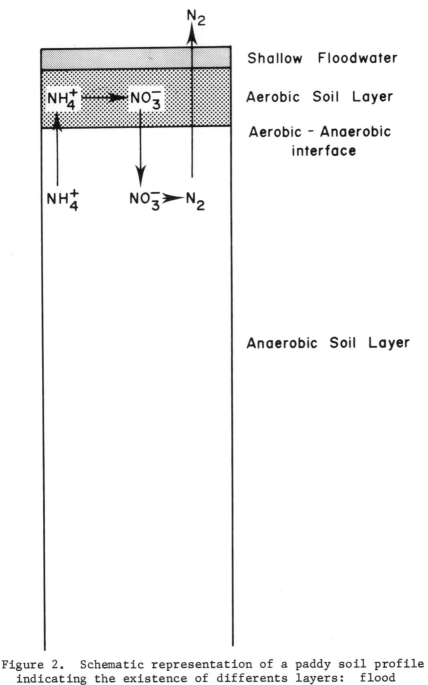

Figure 2.   Schematic representation of a paddy soil profile
   indicating the existence of differents layers: flood
   water, aerobic and aerobic-anaerobic interface, anaerobic
   layer (Reddy *et al.*, 1976).

responsible for the occurrence of at least two more microenvironments in the so-called anaerobic soil layer: the
plant rhizosphere microenvironment and the root litter and
stubble environment.

Flood Water Environment

Microorganisms are present in large numbers in the
water layer of paddy fields. Among them, algae together with
aquatic weeds, play a prominent role by supplying oxygen
through photosynthesis. Diurnal variations in oxygen dissolved
in the water layer are well known and are responsible for
variations in the oxidation-reduction equilibrium of the water
layer itself and also of the soil-water interface. The
concentration of dissolved O_2 was reported to vary from 2 to
18 ppm (Yoshida, 1975).

The role of algae in the water layer, especially as
N_2-fixing organisms, was studied extensively by Venkataraman
(1975) in Asia and recently by Reynaud and Roger (1977) in
Western Africa. Watanabe *et al.*, (1977) showed that when
replacing flooding water and floating algal mass with distilled water, *in situ* N_2 fixation ($C_2H_2-C_2H_4$) was dramatically
reduced in IRRI's paddy fields from 117 mol C_2H_4 per 24 hr
to 19 mol when algae bloomed. At a later stage, the removal
of algae from flooding water did not influence *in situ* N_2
fixation of the paddy soil. Reynaud and Roger (1977) reported
that, in Senegal, growth and N_2-fixing activity of blue-green
algae were dependent upon the rice plant cover. At early
stages, N_2-fixing activity was low, due to the inhibitory
effect of high light intensity (70,000 lux at 1:00 pm).
Later, however, when the rice leaves protected the water
layer against excessive light, blue-green algae could prolifer-
ate and fix N_2 actively. Thus the water layer appears to
constitute a favorable environment for N_2 fixation, but external changes (*e.g.*, light intensity) may alter this
characteristic.

The Aerobic Soil Layer

A thin oxidized layer develops in the upper part of
the paddy soil. Aerobic processes, such as nitrification,
are known to take place in this horizon (Greene, 1960).
Simultaneously denitrification and ammonification may occur
in lower horizons, causing large losses of N_2.

The Anaerobic Soil Layer

The horizon placed beneath the aerobic soil layer is
anaerobic and more or less reduced. This horizon is the site
of anaerobic processes

The Anaerobic Soil Layer

The horizon placed beneath the aerobic soil layer is anaerobic and more or less reduced. This horizon is the site of anaerobic processes, but, contrary to the often-held belief, such processes occur only in favorable microenvironments, especially in sites of accumulation of root litter or stubble.

The Rhizosphere Microenvironment

It has long been known that the microbial population and microbial activity in the rhizosphere differs markedly from that occurring in the soil itself. Since the total surface area of plant roots is much larger than the soil surface occupied by the plants, one can easily predict that the rhizosphere environment should play a major role in the biology of paddy soils. In the root environment rice plants greatly influence different microbial activities, especially N_2 fixation, denitrification, sulfate reduction, methane production.

Among the unique characteristics of the rice root is its ability to facilitate the transfer of oxygen from the foliage to the rhizosphere (Ishizuka, 1971; Luxmoore *et al.*, 1970). Recent tracer studies showed that in the rice rhizosphere the relative partial pressure was still 0.2 for 40-cm long roots, whereas the corresponding pressure for corn and barley was nil (Figure 3). Using a ^{15}N tracer technique,

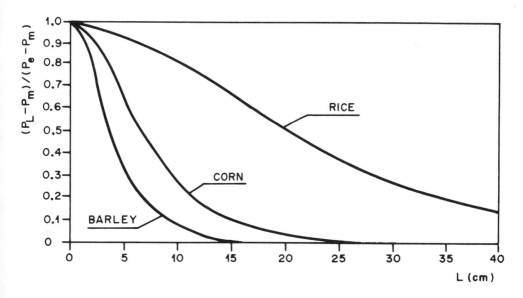

Figure 3. Relative partial pressure of oxygen at the root tip as a function of the root length (L) (Jensen *et al.*, 1967).

Yoshida and Broadbent (1975) confirmed that a similar transfer occurred for atmospheric nitrogen. At the tillering stage,
the rate of atmospheric nitrogen diffusion through the plant
appeared to slow down, whereas this rate was much greater at
the heading or flowering stage. Thus pO_2, pN_2, pCO_2 gradients
around the rice root differ markedly from those observed
around plants growing in drained soils.

N_2 *Fixation*

Rice roots harbor diazotrophic bacteria belonging to
the following genera: *Pseudomonas, Azotomonas, Azotobacter,
Beijerinckia, Flavobacterium, Arthrobacter, Bacillus, Clostridium, Spirillum, Enterobacter* (Balandreau *et al.*, 1976;
Yoshida, 1970). Most of these bacteria are microaerophilic;
anaerobic bacteria are less abundant (Table II).

Table II
MPN Counts (in bacteria/g of dry soil)
of N_2-Fixing in the Rice Rhizosphere[a]

	Microaerophilic	Anaerobic
Nonrhizosphere soil (control)	3,200,000	372,000
Rhizosphere soil	21,900,000	1,350,000
Rhizosphere soil + roots	61,500,000	472,600

[a]Balandreau *et al.*, 1976.

In paddy fields, rhizosphere bacteria together with
nonrhizosphere diazotrophs contribute to N_2 fixation:
(1) blue-green algae, (2) nonsymbiotic saprophytic N_2-fixing
bacteria utilizing organic residues, especially the root litter,
and (3) a water fern, *Azolla*, associated with a blue-green
algae (*Anabaena azollae*). N_2 fixation through rhizosphere
bacteria is difficult to measure *in situ* because of the
interference of the other diazotrophs. However, laboratory
experiments show clearly that, in the absence of blue-green
algae, N_2 fixation is located in the root microenvironment
(Haucke-Pacewiczowa *et al.*, 1970).

Denitrification

According to Woldendorp (1963a, b) denitrification in
the rhizosphere of grasses accounts for losses of 15-37 percent
of fertilizer nitrogen added to the soil. This rhizosphere
phenomenon who showed that the rhizosphere soil exhibited
actual and potential denitrification rates up to four times
those of the nonrhizosphere soil (Figure 4). Exudates

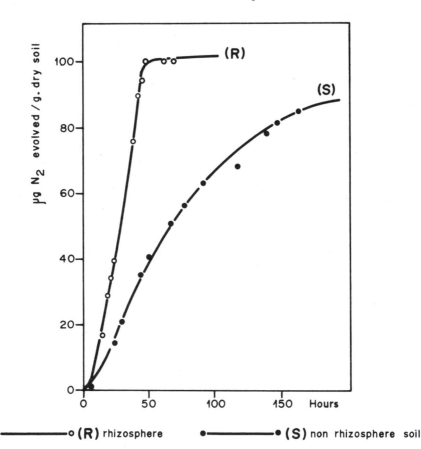

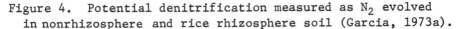

Figure 4. Potential denitrification measured as N_2 evolved
in nonrhizosphere and rice rhizosphere soil (Garcia, 1973a).

are obviously used by denitrifying bacteria as a source of
electrons. R/S ratios of denitrifying bacteria were reported
to vary from 1 to 514, depending on the soil type (Garcia,
1973).

Thus rhizosphere is a preferential site for denitrification. However, the possibility of denitrification in two
other environments must not be overlooked: decaying crop
residues, which act as source of electrons for heterotrophic
denitrifiers and sites containing sulfide or hydrogen, which
are used as electron donors by *Thiobacillus denitrificans*
and *Micrococcus denitrificans*, respectively.

Sulfate Reduction

In spite of the fact that oxygen can diffuse in the
near rhizosphere, sulfate reduction may be observed in the
root zone, provided that the sulfate content of the soil is
high enough (Jacq, 1972; Garcia *et al.*, 1974). Closer

investigations recently showed that, contrary to diazotrophs, sulfate-reducing bacteria were not located in the endorhizo-sphere[1] but were thriving in the rhizoplane (Table III).

Such a conclusion resulted from the comparison of two sets of rice roots. Roots of the first set were thoroughly washed with sterile water, thus eliminating rhizoplane bacteria but unaffecting endorhizosphere bacteria. Roots of the second set were surface sterilized by a 1 percent chloramine T solu-tion that killed rhizoplane bacteria. No sulfate bacteria could survive the surface sterilization treatment, indicating that they were not located in the endorhizosphere and thus were not protected against the sterilizing agent, unlike di-azotrophs (Hamad-Fares *et al.*, 1978).

Methane Production

In Senegalese paddy soils, micropopulation of 10^5-10^7 (per g) methane-producing bacteria were reported by Garcia *et al.* (1974); population sizes in rice rhizosphere are unknown. Raimbault (personal communication) showed that methane evoluation was 6-12 times higher in the rice rhizo-sphere compared with the control without plant. These data provide further evidence of the existence of an anaerobic microhabitat in the rice rhizosphere, which is probably located outside the endorhizosphere.

Root Litter and Stubble Environment

A considerable proportion of the plant biomass is in the form of roots, which are progressively subjected to decay. In spite of the fact that information concerning root decomposition is still scarce (Waid, 1974), one may assume that the input of energy into the soil through decay-ing roots (rhizo-deposition) plays a prominent role through sustaining the different microbial activities at the sites of decomposition.

Probably part of *in situ* N_2 fixation measured in paddy fields should be attributed to diazotrophs associated with decaying rice roots. Rice stubble, which is often ploughed into the soil, may also be a favorable environment of N_2 fixation. Thus Watanabe *et al.*, (1977) found recently that

---

[1]Typically, the rhizosphere can be divided into three areas: (1) the rhizosphere *sensu stricto* (= outer rhizosphere) comprising the region of the soil immediately surrounding the plant roots and the micropopulations living in this zone; (2) the rhizoplane (= root surface) formed by the root surface and the microorganisms living on it; (3) the endorhizosphere (= inner rhizosphere) formed by the root cortical often moribund tissue invaded and colonized by saprophytic soil microorganisms (nonpathogenic host infection).

Table III

Influence of Surface Sterilization of Rice Roots
Upon the Survival of Diazotrophic and Sulfate-Reducing Bacteria[a]

	Number of Bacteria Expressed on a Root Weight Basis		Percentage of 1-cm Long Root Segments Harboring Diazotrophs or Sulfate-Reducing Bacteria	
	Roots Washed[b] with Sterile Water	Roots Surface-Sterilized[c] by Chloramine T	Roots Washed[b] with Sterile Water	Roots Surface-Sterilized[c] by Chloramine T
Diazotrophs	1600×10^6	175×10^6	100	100
Sulfate-Reducing Bacteria	8000	0	67	0

[a] Bauzon and Diem, 1976, personal communication.
[b] Rhizoplane + endorhizosphere bacteria
[c] Endorhizosphere bacteria

N_2 fixation (C_2H_2) was significantly higher in stubble than in the control soil.

Beside diazotrophs, rice debris may harbor active denitrifying bacteria and also active sulfate-reducing or other anaerobic bacteria. As long as such microenvironments act as energy sources for microorganisms, anaerobic hetero-trophs contribute actively to anaerobic transformations of the soil.

THE CONCEPT OF ULTRAMICROENVIRONMENT

Whereas microenvironments are related to a volume commensurate in size with a given organism, ultramicroenviron-ments are characterized by gradations in ions or molecules induced either by organic or inorganic solid particles (*e.g.*, clays, humic compounds) or by other living organisms. Ultramicroenvironments induced by solids, which have already been described by McLaren and Skujins (1968) and Hattori and Hattori (1976) as molecular environments, are not dealt with here.

The concept of ultramicroenvironment induced by other organisms has not yet emerged clearly. Two examples will illustrate this concept. The first example is related to the association occurring between *Rhodopseudomonas capsulatus* and *Azotobacter vinelandii* (Figure 5). In spite of being an anaerobic bacteria, *R. capsulatus* can grow well when associated

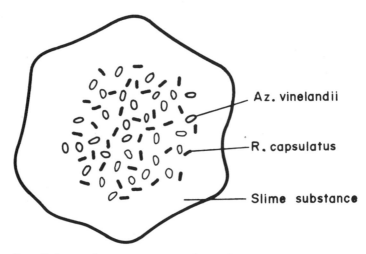

Figure 5. Schematic representation of a mixed culture of *Rhodopseudomonas capsulatus* and *Azotobacter vinelandii* (Okuda *et al.*, 1961).

with *A. vinelandii* even in aerobic environments, the inter-action between *R. capsulatus* and *A. vinelandii* occurring near or on their cell membrane. The second example is that of

Methanobacillus omelianskii, which transforms ethanol and CO_2 into acetate and methane. Actually *M. omelianskii* is a mixed culture of two microorganisms - a methane-producing bacteria (H) and another organism (S) that oxides ethanol into acetate and H_2. The reactions operated by each organisms are:

Organism H: $\quad 4\ H_2 + CO_2 \longrightarrow CH_4 + 2\ H_2O$

Organism S: $\quad \dfrac{2\ CH_3CH_2OH + 2\ H_2O \longrightarrow 2\ CH_3-COOH + 4\ H_2}{2\ CH_3CH_2OH + CO_2 \longrightarrow 2\ CH_3-COOH + CH_4}$

Organisms S does not grow properly on ethanol. But, if both organisms are grown in mixed culture, they grow perfectly well on ethanol and CO_2. The interaction between the organisms consists of an interspecific transfer of H_2 (Bryant *et al.*, 1967; Le Gall, 1977, personal communication). In such a system, organism H induced an environment promoting the oxydation of ethanol into acetate by organism S.

Of course the concept of ultramicroenvironment is not limited to paddy fields and can be extrapolated to other soil types.

VARIATIONS OF MICROBIAL ACTIVITY IN MICROENVIRONMENTS

Such variations were clearly demonstrated in the case of N_2 fixation in the rice rhizosphere. Since diazotrophs thriving in the rhizosphere depend upon the supply of energy by the plant and since this supply depends itself upon the plant photosynthesis and translocation of energy yielding compounds towards the roots, one can predict that N_2 fixation by diazotrophs associated to the root will depend upon the factors governing the plant physiology, especially light intensity and temperature. Actually, *in situ* measurements showed that N_2 fixation fluctuated diurnally, with a midday peak a minimum level during the night (Figure 6). Seasonal variations have been stressed already in this chapter, *e.g.*, variations of N_2 fixation by blue-green algae induced by modification of light intensity reaching the flood water layer.

DIFFUSION OF METABOLIC PRODUCTS ORIGINATING FROM THE THE DIFFERENT MICROENVIRONMENTS

Since each of the microenvironments making up the paddy soil are the sites of different microbial reactions occurring most often simultaneously, different metabolic products accumulate at the same time, creating gradients of concentration around the most active sites. Thus the concentration of NH_4^+-N in the aerobic layer tends to decrease rapidly since NH_4^+-N is oxidized to NO_3^--N by nitrifying bacteria,

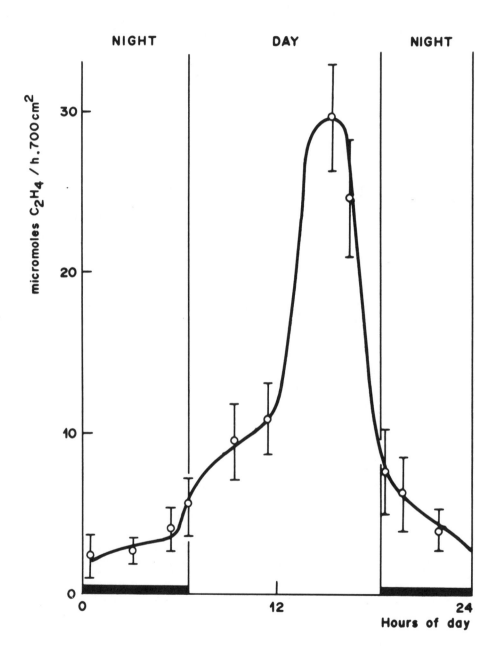

Figure 6.   Diurnal variations of rhizosphere nitrogen fixation
in a rice field assayed by acetylene reduction; calculated
curve observed means and standard errors (shown by limits)
(Balandreau *et al.*, 1974).

which find favorable environmental conditions in this layer. NH_4^+-N, resulting from ammonification of root debris in the anaerobic layer, diffuse upwards into the aerobic layer. This diffusion process is influenced by several factors such as organic matter status of the soil, cation exchange capacity, bulk density, and presence of reduced Fe and Mn (Reddy *et al.*, 1976). Simultaneously NO_3^--N readily diffuses back down into the anaerobic layer and is subsequently denitrified. Losses of N_2 and N_2O through these simultaneous processes are reported to be large (Broadbent and Tusneem, 1971; Yoshida and Padre, 1974). Thus, physical diffusion processes may enhance specific biological reactions, which may be detrimental to the fertility status of the soil, such as denitrification.

REFERENCES

Balandreau, J., C. R. Millier and Y. Dommergues. "Diurnal Variations of Nitrogenase Activity in the Field," *Appl. Microbiol.* 27(662)

Balandreau, J., G. Rinaudo, I. Hamad-Fares and Y. Dommergues. "Nitrogen Fixation in the Rhizosphere of Rice plnat," in *Nitrogen Fixation in the Free Living Microorganisms*, W. D. P. Stewart, ed. (Cambridge: Cambridge University Press, 1975), pp. 57-70.

Balandreau, J. P., G. Rinaudo, M. M. Oumarov and Y. Dommergues. "Asymbiotic N_2 Fixation in Paddy Soils," in *The Proceedings 1st International Symposium on N_2 Fixation*, vol. 2, W. E. Newton and C. J. Nyman, eds. (Pullman: Washington State University Press, 1976), pp. 611-628.

Broadbent, F. E. and M. E. Tusneem. "Losses of Nitrogen From Some Flooded Soils in Tracer Experiments," *Soil Sci. Soc. Amer. Proc.* 35:922-926 (1971).

Bryant, M. P., E. A. Wolin, M. J. Wolin and R. S. Wolfe. "*Methanobacillus omelianskii*, a Symbiotic Association of Two Species of Bacteria," *Arch. Mikrobiol.* 59:20-31 (1967).

Garcia, J. L. "Influence de la rhizosphere du riz sur l'activité dénitrifiante potentielle des sols de rizière du Senegal," *Oecol. Plant* 8:315-323 (1973a).

Garcia, J. L. "Séquence des produits formés au cours de la dénitrification dans les sols de rizières du Senegal," *Ann. Microbiol.* (Institut Pasteur, 1973b) 124B:351-362.

Garcia, J. L.   "Evaluation de la dénitrification dans les rizières par la méthode de reduction de N_2O," *Soil Biol.* 7:251-256 (1975).

Garcia, J. L., M. Raimbault, V. Jacq, G. Rinaudo and P. Roger. "Activités Microbiennes dans les Sols de Rizières du Senegal: Relations avec les Caractéristiques Physico-Chimiques et Influence de la Rhizosphère," *Rev. Ecol. Biol. Sol* 2:169-185 (1974).

Greene, H.   "Paddy Soils and Rice Production," *Nature* 186:511-513 (1960).

Hamad-Fares, I., H. G. Diem, M. Rougier, J. Balandreau and Y. Dommergues. "Colonization of Rice Roots by Diazotroph Bacteria," in *Proceedings of the International Symposium on Environmental Role of Nitrogen Fixing Blue-Green Algae and Asymbiotic Bacteria*, (Stockholm: Uppsala. Bull. Ecol. Res. Comm., in press).

Hattori, T. and R. Hattori. "The Physical Environment in Soil Microbiology: An Attempt to Extend Principles of Microbiology to Soil Microorganisms," *CRC Critical Reviews in Microbiology* mai, 423-461 (1976).

Haucke-Pacewiczowa, T., J. Balandreau and Y. Dommergues. "Fixation Microbienne de 1 Azote dans un Sol Salin Tunisien," *Soil Biol. Biochem.* 2:47-53 (1970).

Ishizuka, Y.   "Physiology of the Rice Plant," *Adv. Agron.* 23:241-315 (1971).

Jacq, V.   "Biological Sulphate-Reduction in the Spermosphere and the Rhizosphere of Rice in Some Acid Sulphate Soils of Senegal," in *Proceedings of the International Symposium Acid Sulphate Soils* (International Institute for Land Reclamation and Improvement (J.L.R.J.), 1972), 18:82-96.

Jensen, C. R., L. H. Stolzy and J. Letey. "Tracers Studies of Oxygen Diffusion Through Roots of Barley, Corn and Rice," *Soil Science* 103:23-29 (1967).

Luxmoore, R. J., L. H. Stolzy and J. Letey. "Oxygen Diffusion in the Soil-Plant System. IV. Oxygen Concentration Profiles, Respiration Rates, and Radial Oxygen Losses Predicted for Rice Roots," *Agron. J.* 62:329-332 (1970).

McLaren, A. D. and J. Skujins. "The Physical Environment of Microorganisms in Soil," in *The Ecology of Soil Bacteria*, Gray and Parkinson, eds., (Liverpool: University Press, 1968), pp. 3-24.

Okuda, A., M. Yamaguchi, M. Kobabayashi and T. Katamaya. "Nitrogen-Fixing Microorganisms in Paddy Soils. VIII. Nitrogen Fixation in Mixed Culture of Photosynthetic Bacteria (*Rhodopseudomonas capsulatus* species) with Other Heterotrophic Bacteria (2) *Soil Sci. Pl. Nutr.* 7:146-151 (1961).

Reddy, K. R., W. H. Patrick and R. E. Phillips. "Ammonium Diffusion as a Factor in Nitrogen Loss from Flooded Soils," *Soil Sci. Soc. Amer. J.* 40:528-533 (1976).

Redman, F. H. and W. H. Patrick, Jr. "Effect of Submergence on Several Biological and Chemical Soil Properties," *Agric. Exper. Station* 592:3-28 (1965).

Reynaud, P. A. and P. A. Roger. "N_2-Fixing Algal Biomass in Senegal Rice Fields," in *International Symposium on Environmental Role of N_2 Fixing Blue-Green Algae and Asymbiotic Bacteria*, Uppsala (in press).

Turner, F. T. and W. H. Patrick. "Chemical Changes in Waterlogged Soils as a Result of Oxygen Depletion,: in *9th International Congress of Soil Science Transactions*, Vol. IV:53-65 (1968).

Venkataraman, G. A. "The Role of Blue-Green Algae in Tropical Rice Cultivation," in *International Biological Programme Nitrogen Fixation by Free-Living Microorganisms*, Vol. 6 W. D. P. Stewart, ed. (Cambridge: Cambridge University Press, 1975), pp. 207-219.

Waid, J. S. "Decomposition of Roots," in *Biology of Plant Litter Decomposition*, Vol. 1 C. H. Dickinson and G. J. F Pugh, eds. (London and New York: Academic Press, 1974), pp. 175-211.

Watanabe, I., K. K. Lee, B. V. Alimagno, M. Sato, D. C. Del Rosario and M. R. de Guzman. "Biological Nitrogen Fixation in Paddy Field Studied by *in situ* Acetylene-Reduction Assays," *IRRI Research Paper Series* 3:1-16 (1977).

Woldendorp, J. W. "The Influence of Living Plants on Denitrification," *Meded. Landbouwhogeschool* 63:1-100 (1963a).

Woldendorp, J. W. "L'influence des Plantes Vivantes sur la Denitrification," *Ann. Institut Pasteur* 105:426-433 (1963b).

Yoshida, T. "Microbial Metabolism of Flooded Soils," in *Soil Biochemistry* Vol. 3 E. A. Paul and A. D. McLaren, eds. (New York: Marcel Dekker, Inc., 1975), pp. 83-115.

Yoshida, T. and F. E. Broadbent. "Movement of Atmospheric Nitrogen in Rice Plants," *Soil Science* 120:288-291 (1975).

Yoshida, T. "Soil Microbiology," in *Annual Report, The International Rice Research Institute,* (Manila: 1970), pp. 47-59.

CHAPTER 36

FREE RADICALS IN HUMIC SUBSTANCES

N. SENESI

Instituto di Chimica Agraria
Facolta di Agraria
Universita di Bari
Bari, Italy

M. SCHNITZER

Soil Research Institute
Agriculture Canada
Ottawa, Ontario, Canada

INTRODUCTION

Humic compounds contain relatively high amounts of stable free radicals of the semiquinone type (Rex, 1960; Steelink and Tollin, 1962; Atherton *et al.*, 1967; Schnitzer and Skinner, 1969; Lisanti *et al.*, 1974). Apparently they are stable over periods of years under air, can survive the geochemical processes of humification and coalification, and may have wide-ranging effects on many reactions that occur in soils. It is likely that the free radicals participate actively as intermediates in oxidation-reduction reactions in terrestrial and aquatic environments. Thus, these compounds play an important role as electron donors and acceptors in nature and so are capable of governing many biogeochemical reactions that occur in these systems.

Recently, Schindler, Williams and Zimmermann (1976) tested by biological methods the ability of microorganisms to utilize humic substances as electron transport systems. They demonstrated that humic substances are capable of catalyzing the transport of electrons from reduced materials to electron acceptors that are normally thermodynamically unfavorable and that they are also capable of enhancing rates of already feasible reactions. So, under environmental conditions, anaerobic microbial populations can utilize humic substances as electron pathways.

It is well known that flavin electron transport involves the formation of semiquinone intermediates. Similarly, humic substances rich in phenolic and quinonoid groups (Ognar and Schnitzer, 1971; Riffaldi and Schnitzer, 1972; Flaig *et al.*, 1975) can mimic flavoproteins by disproportionating as metastable semiquinones (Schindler *et al.*, 1976). Although it appears that semiquinone intermediates are involved in oxidation-reduction processes of humic substances, little is known about the mechanisms that control these reactions in aqueous solutions and their response to many environmental factors, such as changes in pH, exposure to long reaction times, light irradiation and treatments with different oxidants and reductants. We were also interested in discovering whether these reactions were reversible. While some of these effects have been investigated previously (Blois, 1955; Steelink *et al.*, 1963; Lagerkrantz and Yhland, 1963; Steelink, 1964, 1966; Slawinska *et al.*, 1973), the results are often incomplete or contradictory.

Because of the high sensitivity of electron spin resonance (ESR) spectrometry for following minute structural changes involving free radical intermediates, we decided to reexamine and to elaborate on effects caused on humic substances by the above mentioned physical and chemical factors. The humic material employed in this study was a fulvin acid (FA), which was extracted from a Podzol Bh horizon. This material is completely soluble in H_2O at all pH's and is known to occur widely in both terrestrial and aquatic environments.

MATERIALS AND METHODS

Origin and Extraction of FA

The FA originated from the Bh horizon of the Armadale soil, a poorly drained Podzol (4.0% C, pH (H_2O) = 4.0) on Prince Edward Island. Methods used for extraction, fractionation and purification of the FA and for the elementary and functional group analyses of this material have been described (Ogner and Schnitzer, 1971). The purified, moisture and ash-free FA contained 50.9% C, 3.3% H, 0.7% N, 0.3% S and 44.7%O.

One g of FA contained 9.1 meq CO_2H, 3.3 meq phenolic OH, 3.6 meq alcoholic OH, 2.5 meq ketonic C = 0, 0.6 meq quinonoid C = 0 and 0.1 meq OCH_3. The Mn (number-average) molecular weight of the FA was 951 (Hansen and Schnitzer, 1969).

Preparation of FA Solutions

One percent (w/v) FA solutions were prepared by dissolving 100 mg of FA in water and raising the pH from 2 to 10 by the addition of dilute NaOH solution. Solutions at pH > 10 were prepared by dissolving the FA directly in 0.1 N or 1.0 N NaOH. The natural pH of a one percent (w/v) FA solution in H_2O was $\sim$ 2. ESR parameters of these solutions were measured and their variations with time were observed.

Reduction of FA Solutions

Reductant: FA molar ratios of 1:1 were employed. Thus, 4.5 mg solid $SnCl_2 \cdot 2H_2O$ was added to 2 ml one percent (w/v) FA solutions at pH 2, 7 and > 13. After mixing, only the system at pH 2 appeared heterogeneous, but all other systems were homogeneous. For the reduction by $NaBH_4$, 100 mg FA was dissolved in 10 ml of 1 N NaOH, which was 0.01 M in $NaBH_4$ (pH $\gtrsim$ 13). ESR spectra of these systems were recorded during the reduction.

Oxidation of FA Solutions

Molar oxidant: FA ratios were 1:1. In the case of H_2O_2, 0.7 ml 0.5% H_2O was added to solutions prepared by dissolving separate 100 mg portions of FA in either 9 ml 1.0 N NaOH (pH > 13), 9 ml distilled water only (pH 2) or 9 ml distilled water adjusted to pH 7 and 10 with 1.0 N NaOH. Oxidations with Ag_2O and $NaIO_4$ were performed on separate volumes of one percent (w/v) FA solutions at pH > 13 and 2, respectively, by adding 23.2 mg Ag_2O or 21.5 mg $NaIO_4$. During oxidations ESR spectra were recorded as a function of reaction time.

Irradiation Experiments

Untreated, oxidized and reduced FA solutions at various pH levels were irradiated from 5 to 50 min *in situ* through the slots in the ESR spectrometer cavity with a 500-W tungsten lamp mounted 20 cm from the cavity. Irradiation increased the temperature in the cavity by about 5-8°C. ESR spectra were recorded during and after irradiation.

Reversibility Experiments

After 4 hr of oxidation (as previously described), separate volumes of each oxidized solution were treated at pH 2, 7 and > 13 with the following reductants: (a) $SnCl_2$, (b) $Na_2S_2O_4$ and (c) $NaBH_4$ (at pH > 13 only). Molar reductant: FA Ratios employed were 1:1. ESR spectra were recorded during reduction until maximum signal intensities were attained. Each reduced solution was then reoxidized with the same reagent employed for the first oxidation (at a molar ratio of 1:1) and ESR spectra were recorded after 10 min of reaction.

Electron Spin Resonance (ESR) Measurements

ESR spectra were recorded at room temperature on a
Varian Associates E-3 spectrometer, employing 100-KHz modu-
lation and a nominal operation frequency of 9.5 GHz. Spin
concentrations, line widths and G-values were computed from
the spectra according to Riffaldi and Schnitzer (1972).

RESULTS

Effects of pH and Time on ESR Parameters

ESR spectra of FA solutions at various pH's consisted
of single, symmetrical lines devoid of hyperfine splitting.
ESR parameters of these solutions are presented in Table I.
FA in solutions at acid and neutral pH levels showed free
radical concentrations and G-values that were very similar
to those of solid FA (see note on Table I), but line widths
were considerably narrower in solutions. Spin concentrations
increased markedly with increase in pH; similarly, G-values
tended to increase with pH, but line widths were narrower
at high than at low pH, with solutions at pH > 9 showing the
narrowest line widths (2.0 - 2.2 G).
When the initial pH of the FA solution was < 7.0,
free radical concentrations decreased only very slowly with
time and became constant within 4 hr after dissolution; the
pH remained constant for up to six days. When FA was dissolved in 0.1 N NaOH to give a one percent (w/v) solution,
the pH, immediately after dissolution, was 11.70. After
4 hr, the pH dropped to 10.5, and to 9.0 after six days. At
the same time the free radical content of the FA solution
(Figure 1) first tended to increase, attaining after 20 min
a 9X increase over the value of solid FA, then decreased
gradually. This was followed by a more rapid decrease between
120 and 240 min, reaching a value that was only 25 percent
greater than that of solid FA. Beyond that time, the decrease
in free radical concentrations was relatively slow, declining
to 0.40 x 10^{17} spins/G (33 percent of the value for solid FA)
after six days. During the first 4 hr of the experiment,
G-values remained relatively constant, while line widths
narrowed sharply (from 2.0 to 1.4 G) between 120 and 240 min.
When 10 mg of FA was dissolved in 1 ml of 1.0 N NaOH,
the initial pH was > 13, and remained constant during the
experiment. The free radical content of this solution as
a function of time (Figure 1) followed the same general trend
as that previously described for FA dissolved in 0.1 N NaOH,
except that it attained a maximum value after 60 min (16X the
value of solid FA). This was followed by a rapid decrease
between 60 and 210 min (to a value that was 40 percent greater
than that of solid FA). During the 4-hr experiment, G-values
and line widths remained practically constant.

Table I

ESR Parameters of FA Solutions at Various pH Levels

ESR Parameters[a]		pH							
		2.0	5.0	7.0	8.0	9.0	10.6	11.6	12.5
Spins x 10^{17}/G	Untreated	1.44	1.50	1.56	1.51	1.69	2.19	2.56	12.06
	Irradiated	1.69	1.75	1.82	1.97	2.66	7.77	15.43	17.20
Line Width (G)	Untreated	2.5	3.0	3.0	2.5	2.0	2.0	2.1	212
	Irradiated	2.5	3.2	3.0	2.5	2.0	2.0	2.2	2.2
G-Values	Untreated	2.0038	2.0037	2.0039	2.0040	2.0044	2.0044	2.0044	2.0045
	Irradiated	2.0038	2.0037	2.0039	2.0042	2.0044	2.0044	2.0043	2.0044

[a]ESR parameters for solid FA were: 1.50 spins x 10^{17} G; line width = 4.5 G; G-value = 2.0040.

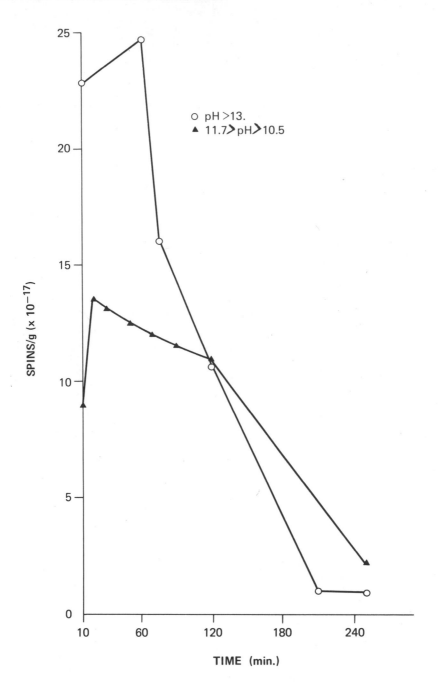

Figure 1.   Changes of spin concentrations with time in FA solutions at alkaline pH's.

Effects of Reduction of ESR Parameters

The following two reducing agents were used for this purpose: (a) $NaBH_4$ at pH > 13 and (b) $SnCl_2$ at several pH's. Reductions were performed on FA solutions that had been allowed to stand at room temperature for many hours. ESR parameters for FA solution during reduction at various pH's are presented in Table II. Very strong increases in free radical concentrations were generally observed within 10 to 40 min. However, in the case of reduction with $SnCl_2$ at low pH maximum spin, concentrations were attained only after 21 hr. The observed maxima were between 40X and 108X the initial values, and between 17X and 91X the value of solid FA. These maxima were followed by gradual decreases. Values very close to untreated FA solutions and much lower than the value for solid FA were reached within less than two days in FA solutions at alkaline pH but still later at neutral and acidic pH's. Gravity-values and line widths remained more or less constant when reductions were done at high pH, but G-values decreased at pH 7 and 2 and line widths broadened on reduction at pH 2, probably because of fine colloidal aggregates that appeared in the system.

Effects of Oxidation on ESR Parameters

The oxidants that we employed were: (a) H_2O_2 at all pH's, (b) Ag_2O at pH > 13 and (c) $NaIO_4$ at pH 2. Figure 2 illustrates changes in spin concentrations with time during oxidation of FA with H_2O_2 at various pH's. The curves for the oxidation of FA with Ag_2O at pH > 13 oxidation at the same pH's. Very strong decreases in free radical concentrations were observed, with the highest rates at the highest pH's. In all cases, minimum values of spin concentrations were attained within 4 hr and these were generally much lower than the value for solid FA (Table III, second line). No significant changes in G-values (2.0047 – 2.0040), but changes in line widths were observed. Hyperfine splitting, consisting of a unsymmetrical 3-line signal, was observed when FA was oxidized with Ag_2O at pH > 13 and with H_2O_2 at pH 7 (these results are discussed in detail by Senesi and Schnitzer, 1976).

Effects of Irradiation with White Light on ESR Parameters

Table I shows ESR parameters for FA solutions at various pH's irradiated for 5 min with white light of 500-W intensity. Increases in free radical concentrations were observed, especially at pH's > 9.0; the maximum relative increase (6X the value for the original FA solution at the same pH and 10X that of solid FA) occurred at pH 11.6.

Table II
ESR Parameters for FA Solutions During Reduction

Time (min)	Spins × 10^17/G			Line Widths (G)				G-Values			
	NaBH₄ pH > 13	SnCl₂ pH = 7[a]	pH = 2[b]	NaBH₄ pH > 13	SnCl₂ pH > 13	SnCl₂ pH = 7[a]	pH = 2[b]	NaBH₄ pH > 13	SnCl₂ pH > 13	SnCl₂ pH = 7[a]	pH = 2[b]
0	0.46	1.27	1.49	2.0	2.0	3.0	2.5	2.0042	2.0042	2.0030	2.0038
10	30.60	50.65	82.58	2.2	n.d.	3.3	5.3	2.0045	n.d.	2.0023	2.0027
40	29.83	37.37	88.65	2.1	2.1	3.0	5.3	2.0042	2.0041	2.0021	2.0025
960	n.d.	7.62	90.73	n.d.	2.2	3.0	5.1	n.d.	2.0040	2.0028	2.0027
1260	n.d.	4.63	137.03	n.d.	2.2	3.0	5.0	n.d.	2.0039	2.0027	2.0022
1440	0.40	3.73	136.20	2.2	2.2	3.0	5.0	2.0042	2.0040	2.0026	2.0027
2610	0.40	2.31	94.68	2.2	2.2	3.0	5.0	2.0039	2.0039	2.0028	2.0026

[a] pH drops to 5.0 after addition of SnCl₂.
[b] pH drops to 1.7 after addition of SnCl₂.
n.d. not determined.

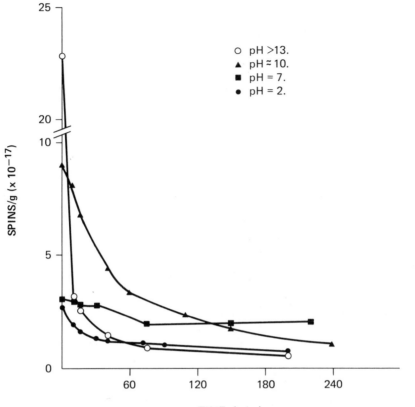

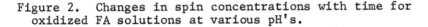

Figure 2.   Changes in spin concentrations with time for
   oxidized FA solutions at various pH's.

FA solutions at pH's 2, 7 and 10 were irradiated for
50 min and then the light was turned off. The relative spin
concentrations of these solutions during irradiation as a
function of time are presented in Figure 3. The free radical
content attained maximum values after 5 min of irradiation
in all cases; this was followed by gradual decreases. The
highest increase was shown by a FA solution at pH ∿ 10 and
the increase was sustained during 50 min of irradiation.
When the light was switched off, free radical concentrations
returned rapidly to values that were close to those prior to
irradiation for FA solutions at pH's ∿ 10 and 2, and to a
lower value for the FA solution at pH 7.
   Irradiation of the oxidized FA solutions was more
effective in increasing free radical concentrations at high
pH than at neutrality and under acidic conditions (Table III,
third line). Irradiation of reduced FA solutions after they

Table III
Effects of Oxidation and Irradiation on ESR Parameters

FA	pH > 13	H_2O_2 pH ~ 10	pH 7	pH 2	Ag_2O pH > 13	$NaIO_4$ pH 2
Untreated	22.86	9.03	2.99	2.69	22.86	2.69
Oxidized[a]	0.46	1.00	2.09	0.75	0.40	1.20
Irradiated[b]	21.26	2.86	0.75	1.49	15.41	1.79
Reduced[c] with						
$NaBH_4$	30.70	7.71	–	–	31.36	–
$SnCl_2$	24.85	13.02	33.22	54.11	21.79	n.d.
$Na_2S_2O_4$	30.04	1.00	11.60	7.62	13.02	4.04
Reoxidized[d] after						
$NaBH_4$	0.90	0.95	–	–	0.73	–
$SnCl_2$	0.86	0.87	3.59	1.94	2.79	n.d.
$Na_2S_2O_4$	0.94	0.83	0.53	0.90	1.99	1.12

[a] Minimum values attained after 4 hr oxidation.
[b] For 5 min, after 4 hr oxidation.
[c] Maximum values attained after 4 hr oxidation.
[d] After the maximum values were attained on reduction.
n.d. not determined.

had attained maximum free radical concentrations did not
increase free radical content any further. Throughout the
irradiation experiments G-values and line widths remained
practically unaffected.

Reversibility of Oxidation-Reduction Reactions

To examine the possible reversibility of the oxidation-
reduction reactions involving FA free radicals, we treated
FA with the following sequential reactions: oxidation →
reduction → oxidation. Effects of the different consecutive
treatments on the free radical concentrations of FA solutions
are summarized in Table III. In all cases reduction of oxidized FA solutions led to strong increases in free radicals.
Maximum free radical concentrations were obtained at different
rates, depending on the type of reducing agent and the pH of
the system (times ranged between 10 and 75 min). The maxima
were not sustained for a long time, but the free radical content decreased gradually, attaining values close to those
before reduction within less than two days. Reoxidation of
the reduced FA solutions after the maximum free radical content had been attained brought about a rapid decay of free
radicals to values before reduction.

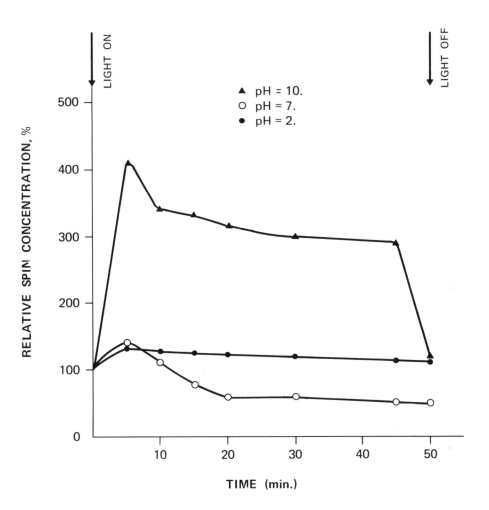

Figure 3. Effect of irradiation time on spin concentrations of FA solutions at various pH's.

DISCUSSION

An inspection of the results reported herein shows that over short periods of time, the dissolution of FA in aqueous solutions increases free radical concentrations to values that are higher than that of solid FA. The extent of the increase is a direct function of increasing the pH. The increases in free radicals are not sustained for long, but the free radicals tend to decrease at rates that are inversely related to the initial increases. At alkaline pH the initial increase in free radicals is very strong but they decay rapidly, whereas at neutral and acidic pH, when the initial increases

are relatively smaller, rates of decay are also slower. In
all cases, after several days of standing at room temperature,
spin concentrations of FA in aqueous solutions are lower than
those for solid FA. Irradiation with visible light and reduction also strongly increase free radical concentrations
of FA solutions. Again, after these treatments, increases
in free radicals are followed by relatively fast decreases
to values that are below the value for solid FA. In all
instances, treatments with oxidants have the effect of rapidly
reducing the high free radical contents of freshly prepared
FA solutions at different pH's. Rates of decay of the oxidized
FA solutions are much higher than those of untreated ones.

So we can observe the presence in FA solutions of at
least two types of free radicals: (a) permanent ones, which
persist over long periods of time, such as those in solid FA,
and (b) transient ones, which are generated by raising the
pH, reduction or irradiation and which are not stable and
do not resist oxidation. Moreover, some free radicals that
are stable in the solid state are lost in solution after
short periods of time. This may be due to a higher possibility
for irreversible dimerization and/or polymerization reactions
in the solution phase (Senesi and Schnitzer, 1976). G-values
observed before and after the various treatments range between
2.0037 and 2.0045 and are similar for permanent and transient
radicals, indicating that both are substituted semiquinones
(Blois *et al.*, 1961). Line widths in solutions are narrower
than those measured on suspensions or on powdered samples
because of more rapid tumbling of molecules in liquid phase
and greater freedom of rotation (Ingram, 1958; Senesi and
Schnitzer, 1976).

Oxidation-reduction reactions of FA in aqueous solutions involve mainly reversibly transient free radicals.
This suggests that: (a) these reactions proceed by two one-
electron steps via semiquinone radical intermediates, (b)
these intermediates are not sufficiently long-lived under
oxidative conditions but sufficiently stable under reducing
conditions or after irradiation, (c) that FA, as well as
other humic substances, can act as electron donor-acceptor
systems (Ziechmann, 1972) and so participate in oxidation-
reduction reactions with transition metals and biological
systems in the soil and water environment (Schindler *et al.*,
1976), and (d) that this participation is high at pH 7, in
waterlogged or poorly-drained soils where reducing conditions
prevail, and near surfaces of soils and waters exposed to
sunlight.

Steelink (1966) rationalizes free radical reactions
in humic substances in terms of intermolecular charge transfer
systems consisting of quinone and hydroquinone structures.
In Figure 4 we propose a scheme that explains the observed
reversibility of the oxidation-reduction reactions, their

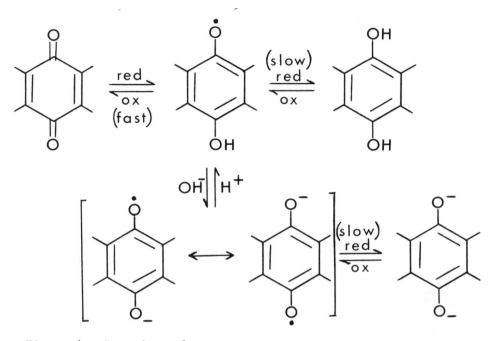

Figure 4.   Reaction scheme.

homolytic mechanism, and the relative stability of the transient free radical intermediates under different redox, pH and
irradiation conditions to which FA can be subjected in terrestrial and aquatic microenvironments.

REFERENCES

Atherton, N. M., P. A. Cranwell, A. J. Floyd and R. D. Haworth.
  "Humic-Acid-I: ESR Spectra of Humic Acids," *Tetrahedron*
  23:1653-1667 (1967).

Blois, S. "A Note on the Radical Formation in Biologically
  Occurring Quinones," *Biochim. Biophys. Acta* 18:165 (1955).

Blois, M. S., Jr., H. W. Brown and J. E. Maling. "Precision
  G-Value Measurements on Free Radicals of Biological Interest,"
  in *Free Radicals in Biological Systems*, M. S. Jr. Blois,
  H. W. Brown, R. M. Lemmon, R. O. Lindblom and M. Weissbluth,
  eds. (New York, London: Academic Press, 1961), pp. 117-131.

Flaig, W., H. Beutelspacher and E. Rietz. "Chemical Composi-
  tion and Physical Properties of Humic Substances," in
  Soil Components, vol. 1, J. E. Gieseling, ed. (New York:
  Springer-Verlag, Inc, 1975), pp. 1-211.

Hansen, E. H. and M. Schnitzer. "Molecular Weight Measurements of Polycarboxylic Acids in Water by Vapor Pressure
  Osmometry," *Anal. Chim. Acta* 46:247-254 (1969).

Ingram, D. J. E. *Free Radicals as Studied by Electron Spin Resonance* (London: Butterworth Scientific Publication, 1958), pp. 102-159.

Lagercrantz, C. M. and M. Yhland. "Photo-Induced Free Radical Reactions in the Solutions of Some Tars and Humic Acids," *Acta Chem. Scand.* 17:1299-1306 (1963).

Lisanti, L. E., C. Testini and N. Senesi. "Research on the Paramagnetic Properties of Humic Compounds. VI. Identification of Free Radicals by Electron Paramagnetic Resonance (EPR) Spectrometry," *Agrochimica* (in press).

Ogner, J. and M. Schnitzer. "Chemistry of Fulvic Acid, A Soil Humic Fraction and Its Relation to Lignin," *Can. J. Chem.* 49:1053-1063 (1971).

Rex. R. W. "Electron Paramagnetic Resonance Studies of Stable Free Radicals in Lignins and Humic Acids'," *Nature* 188:1185-1186 (1960).

Riffaldi, R. and M. Schnitzer. "Electron Spin Resonance of Humic Substances," *Soil Sci. Soc. Am. Proc.* 36:301-305 (1972).

Schindler, J. E., D. J. Williams and A. P. Zimmerman. "Investigation of Extracellular Electron Transport by Humic Acids," in *Environmental Biogeochemistry*, Vol. 1, J. O. Nriagu, ed. (Ann Arbor: Ann Arbor Science, Inc., 1976), 109-115.

Schnitzer, M. and S. I, M. Skinner. "Free Radicals in Soil Humic Compounds," *Soil Sci.* 108:383-390 (1969).

Senesi, N. and M. Schnitzer. "Effects of pH, Reaction Time, Chemical Reduction and Irradiation of ESR Spectra of Fulvic Acid," *Soil Sci.* (in press).

Senesi, N., Y. Chen and M. Schnitzer. "Hyperfine Splitting in ESR Spectra of Fulvic Acid," *Soil Biol. Biochem.* (submitted for publication).

Slawinska, D., J. Slawinski and T. Sarna. "The Effect of Light on the ESR Spectra of Humic Acids," *J. Soil Sci.* 26:93-99 (1975).

Steelink, C. and G. Tollin. "Stable Free Radicals in Soil Humic Acid," *Biochim. Biophys. Acta* 59:25-34 (1962).

Steelink, C. "Free Radical Studies in Lignin, Lignin Degradation Products and Soil Humic Acids," *Geochim. Cosmoch. Acta* 28:1615-1622 (1964).

Steelink, C. "Electron Paramagnetic Resonance Studies of Humic Acid and Related Model Compounds," in *Coal Geochemistry. Advances in Chemistry* series No. 55, *Am. Chem. Soc.* pp. 80-90 (1966).

Tollin, G., T. Reid and C. Steelink. "Structure of Humic Acid. IV Electron Paramegnetic Resonance Studies," *Biochim. Biophys. Acta* 66:444-447 (1963).

CHAPTER 37

MECHANISMS OF REDUCTIVE TRANSFORMATIONS
IN THE ANAEROBIC MICROENVIRONMENT
OF HYDROMORPHIC SOILS

J. C. G. OTTOW
J. C. MUNCH

Institut für Bodenkunde und Standortslehre
Universität Hohenheim
Stuttgart-Hohenheim
Emil Wolff Strasse 27, Federal Republic of Germany

INTRODUCTION

If a soil is submerged or flooded, a number of characteristic global impacts proceed in a definite sequence. As
a consequence of the blocked gas exchange (O_2, CO_2), the
trapped oxygen is used by respiration within a few hours and
reduction processes caused by microorganisms will soon follow.
The rates of these reductive transformations are different
from soil to soil and depend largely of the nature and amount
of organic matter as well as on the type and amount of inorganic oxidized compounds such as nitrate, Mn(IV)- and
Fe(III)-compounds or sulfate (Ottow and Glathe, 1973).
Generally, the gross sequence of reductive transformations
proceeds in the succession $NO_3^- \rightarrow MnO_2 \rightarrow Fe(OH)_3 \rightarrow SO_4^{2-} \rightarrow$
CO_2 and results in the subsequent production of N_2 and N_2O
(denitrification), Mn(II) and Fe(II)-compounds, blackening
(FeS) and methane. Simultaneously, the soil pH and Eh drop
rapidly (the latter to -450 mV), usually together with a
transient increase in organic acids (Ponnamperuma, 1972;
Yoshida, 1975).
From a thermodynamic viewpoint (Table I), it seems as
if the overall reduction sequence is governed by the pH-Eh
status of the redox systems present in the soil. However,
if these transformations are determined by the pH-Eh level
established, then nitrate should be reduced at any time and
site *before* the reduction of iron(III) oxides will start.
Such a statement is based on the assumption that the reductive
transformations in question are unspecific and primarily of
chemical origin (Patrick and Turner, 1968; Gotoh and Patrick,

Table I
Thermodynamics (pH-Eh Relationships)
for some Redox Systems Operating in Flooded Soils[a]

Redox Systems	Eh (mV) at pH 6	pH 7
$NO_3^- + H_2O + 2e \longrightarrow NO_2^- + 2 H_2O$	+361	+420
$MnO_2 + 4H + 2e \longrightarrow Mn^{2+} + 2 H_2O$	+562	+446
$Fe(OH)_3 + e \longrightarrow Fe^{2+} + 3 OH^-$	+ 98	-108

[a]Data calculated from Ponnamperuma (1972) and Gotoh (1974).

1974). Preliminary studies suggest that such reductions are specific processes, in which nitrate, Mn(II)- and FE(III)-oxides act as electron acceptors for substrate specific reductases, not present in all soil microorganisms (Ottow, 1970; Hammann and Ottow, 19740. This latter view opens some remarkable aspects for the microenvironment of an anaerobic, flooded soil and includes the following postulation:

1. Nitrate will not be reduced *before* Mn(IV)- or Fe(III)-oxides at complete anaerobic sites if the population is free from nitrate reductase but equipped with Mn(IV)- and FE(III)-reductases.

2. Nitrate will prevent Mn(IV)- and Fe(III)-reduction only if the population is capable of producing nitrate-reductase as well as the other specific reductases.

MATERIALS AND METHODS

Bacteria Selected

For model experiments, two different bacteria capable of growing under anaerobic conditions were selected. *Bacillus polymyxa* S 55 was equipped with nitrate reductase (nit⁺) whereas *Clostridium butyricum* S 22a failed to produce this enzyme (Nit⁻). Both strains are capable of fixing nitrogen and reducing considerable amounts of iron. They were isolated from gleyed soils as described previously (Hammann and Ottow, 1976). Both strains were cultivated on a glucose-mineral salt agar (Munch and Ottow, 1977) to obtain spore suspensions (Hammann and Ottow, 1974) for the inoculation of the boiling tubes (150 x 25 mm) that contained 25 ml of the same broth (pH 7.0) plus 0.5 g Fe_2O_3 powder (Merck, sieved to 150-63 um). Other tubes were supplemented with KNO_3 (0.25 mM) and/or 0.1 g MnO_2^- powder (Merck). Enough parallels were prepared to evaluate three tubes per trial.

Chemical Analysis

The inoculated tubes were incubated (at 30° C) anaero-
bically (N_2/CO_2 = 9/1) and analyzed daily (during the first
five days) as well as after five, seven and 14 days on the
following parameters. Mn(II) and Fe(II): Fe(II) in solution
was determined colorimetrically with 2,2-dipyridyl (Hammann
and Ottow, 1974) whereas Mn(II) was measured by AAS (SP 90
A Unicam). The disappearance of nitrate was followed by the
salicylated method (Schlichting and Blume, 1966) that of
glucose with the anthron reagent (Mokrasch, 1954).

Determination of pH and EH

Every tube taken from the incubator was closed (screw
cap), homogenized (vortex mixer) and the pH and Ex measured
using a combined glass–silver chloride–Pt electrode (Ingold,
Type 405) and a Digi 510 pH–meter (WTW) as a recording instrument. The Eh was calculated from the Ex measured by addition
of 207 mV (30° C) and used to determine the rH value (rH =
negative logarithm of the partial pressure of gaseous hydrogen) according to the formula

$$rH = \frac{Eh \ (mv)}{29} + 2 \times pH$$

An rH of 0 corresponds to complete reducing conditions, and
a value of 42.6 indicates entirely oxidizing conditions
(Jacob, 1970). The use of the rH (reduction intensity)
allows the comparison of the redox state within different
systems.

RESULTS

In Tables II (*B. polymyxa*) and III (*C. butyricum*) the
development of Fe(II) in the presence of KNO_3 and/or MnO_2
is compared to the iron reduced under standard conditions.
In order to indicate and compare the state of reduction, the
rH–values are included in each table. The data presented
are those reflecting characteristic stages in the development
of the parameters measured. A complete development of the
various parameters will be published elsewhere in detail
(Munch and Ottow, 1977). The uninoculated controls did not
change significantly throughout the incubation period. In
these controls Fe(II) varied between 1–2 ppm and the Mn(II)
concentration between 10–15 ppm; the production of these
compounds is probably caused by the reducing activity of
glucose.

Table II
Influence of Nitrate, and/or MnO_2
on the pH, rH, Glucose Utilization and Iron-Reduction
in the Culture of *Bacillus polymyxa* (nit^+)
under Complete Anaerobic Conditions

Days	Variables	pH	rH	Fe^{2+} (ppm)	Mn^{2+} (ppm)	NO_3 (mg/ml)	Glucose (mg/ml)
Start	Fe_2O_3	6.5	19.4	0.5	–	–	20.0
	" + KNO_3	6.6	17.8	0.5	–	1.01 ·	20.0
	" + MnO_2	6.5	22.5	1.0	10	–	20.0
	" + both	6.7	21.7	1.8	3	1.01	20.0
2	Fe_2O_3	5.6	0.2	25.7	–	–	17.5
	" + KNO_3	5.8	17.8	2.0	–	0.005	17.2
	" + MnO_2	5.7	3.3	1.9	102	–	18.2
	" + both	6.8	6.1	1.8	73	0	17.6
3.25^a	Fe_2O_3+KNO_3	6.0	0.5	12.0			15.6
						–	
5	Fe_2O_3	6.2	3.5	81.5	–	–	6.2
	" + KNO_3	6.2	4.8	34.0	–	0	10.4
	" + MnO_2	5.8	3.1	2.0	233	–	16.3
	" + both	6.2	4.8	1.8	127	0	11.8
7	Fe_2O_3	6.7	5.7	71.5	–	–	0
	" + KNO_3	6.9	8.5	46.3	–	0	3.7
	" + MnO_2	5.8	2.9	2.5	240	–	14.4
	" + both	6.9	7.6	1.8	60	0	4.0
14	Fe_2O_3	7.3	13.7	48.0	–	–	0
	" + KNO_3	7.4	13.9	30.0	–	0	0
	" + MnO_2	7.3	11.4	2.8	100	–	0.6
	" + both	7.4	15.5	1.8	17	0	0.2

[a] At this time the rH reached nearly complete reducing
conditions.

Table III
Influence of Nitrate and/or MnO_2
on the pH, rH, Glucose Utilization and Iron-Reduction
in the Culture of *Clostridium butyricum* (nit⁻)
under Complete Anaerobic Conditions

Days	Variables	pH	rH	Fe^{2+} (ppm)	Mn^{2+} (ppm)	NO_3 (mg/ml)	Glucose (mg/ml)
Start	Fe_2O_3	6.5	18.9	0.9	–	–	20.0
	" + KNO_3	6.5	18.9	0.9	–	1.01	20.0
	" + MnO_2	6.5	22.8	0.5	13	–	20.0
	" + both	6.5	24.2	1.2	10	1.01	20.0
2	Fe_2O_3	5.0	0	30.0	–	–	13.5
	" + KNO_3	5.2	0	21.5	–	1.01	12.5
	" + MnO_2	6.3	17.9	0.5	29	–	17.9
	" + both	5.7	17.9	1.2	18	1.01	18.7
3.5^a	Fe_2O_3+ MnO_2	5.1	0.5	12.0	108	–	11.2
	"+KNO_3+MnO_2	4.9	0.4	4.0	93	1.01	11.2
5	Fe_2O_3	4.6	10.5	425.0	–	–	1.2
	" + KNO_3	4.7	6.1	200.0	–	1.01	2.8
	" + MnO_2	4.6	4.3	23.0	189	–	2.6
	" + both	4.6	4.4	8.0	173	1.01	3.8
7	Fe_2O_3	4.6	16.6	377.0	–	–	1.1
	" + KNO_3	4.5	16.4	432.0	–	1.01	1.6
	" + MnO_2	4.6	13.8	27.0	258	–	2.2
	" + both	4.6	14.7	12.0	238	1.01	3.5
14	Fe_2O_3	4.6	16.5	377.0	–	–	0.6
	" + KNO_3	4.5	16.4	338.0	–	1.01	0.9
	" + MnO_2	4.5	20.8	25.0	326	–	0.8
	" + both	4.6	21.9	9.0	279	1.01	3.3

[a]At this time the rH reached its minimum.

From Tables II and III the following results may be drawn:

1. rH-development: Both with *B. polymyxa* (nit$^+$) as well as with *C. butyricum* (nit$^-$) the redox-intensity rH decreased rapidly, reaching nearly completely reducing conditions within 2-3 days. In the presence of KNO_3 and/or MnO_2, only a delay in the decrease of the rH, both with *B. polymyxa* as well as with *C. butyricum*, was recorded.

2. Effect of nitrate: The addition of KNO_3 to the medium suppressed the iron-reduction with the nit$^+$ *B. polymyxa* (Table II), but failed to affect it with the nit$^-$ strain of *C. butyricum* (Table III), although *nearly complete reducing condition were established in the culture.*

3. MnO_2: If MnO_2-powder were supplemented to the medium, the amount of dissolved Fe(II), both with the nit$^+$ and nit$^-$ strains, was diminished significantly.

4. MnO_2 + KNO_3: In the presence of both compounds, the reduction of MnO_2 was lowered only if the organism were equipped with nitrate reductase (compare Table II with Table III). The presence of MnO_2 suppressed the amount of reduced iron independently from the occurrence of the enzyme nitrate reductase.

It is appraent from these experiments that the successive reduction of nitrate, manganese and iron is not governed by a decrease in rH (from approximately 19 to nearly 0), but by the activity of specific reductases. In the case of a nit$^+$ bacterium such as *B. polymyxa*, at least two different reductases are involved, one of which seems to be the enzyme nitrate reductase. The other mechanism(s), also responsible for the reduction of Mn(IV) and Fe(III) by nit-organisms such as *C. butyricum* and other clostridia, is specific for Mn(IV) and Fe (III), but not for NO_3^-.

DISCUSSION

If nitrate and MnO_2 are incorporated into a flooded soil, the formation of Fe(II) remains suppressed until the added compounds are reduced (Yuan and Ponnamperuma, 1964; Ponnamperuma *et al.*, 1966; Takai and Kamura, 1966). These observations have supported the view of an indirect, chemical reduction mechanism caused by an unspecific microbial activity and governed by the thermodynamics of the redox systems involved. Although there is no doubt on the chemical transformation of nitrate, MnO_2 or Fe(III)-oxides by organic compounds

(Schnitzer and Skinner, 1967; Szilagyi, 1971), these processes should be disregarded with respect to the activity of certain soil bacteria. Even the establishment of extremely reducing conditions in the soil environment is not sufficient to cause such intensive transformations as have been shown by the results presented. Thus, nitrate suppresses the reduction of Mn(IV)- and/or Fe(III)-oxides only if the reducing organism is equipped with nitrate reductase (*B. polymyxa*) and it is not attacked under complete anaerobic and reducing conditions in the absence of this enzyme (*C. butyricum*).

These results claim the involvement of at least two different and specific reductases, one of which seems to be the nitrate reductase. Further, these reductases are distributed over two types of metabolically different soil populations, the facultative anaerobic nitrate-reducing and denitrifying community (Ottow and Ottow, 1970) and the obligate anaerobic fermenting clostridia (Ottow, 1971). In soil, these organisms are scattered through different microenvironments and various minute habitats.

If such an aerobic soil becomes flooded, the aerobic and facultative anaerobic microorganisms in each habitat increase immediately and are followed by the anaerobic clostridia when complete anaerobic conditions have been established (Takai and Kamura, 1966). The first group of organisms uses nitrate, Mn(IV)- and Fe(III)-oxides as alternative electron acceptors (anaerobic respiration) in the sequence listed, simply because nitrate and MnO_2 require less energy for reduction as Fe(III)-oxides. This is an economical rather than chemical viewpoint. At complete anaerobic conditions, the clostridial population increases and probably switches to Mn(IV) and Fe(III) as electron acceptors as soon as oxidized organic compounds become limited. We believe this characteristic succession in metabolic behavior rather than a lowered redox potential to be responsible for the characteristic successive reduction processes observed in waterlogged soils.

ACKNOWLEDGMENT

This research is supported by a research grant of the Deutsche Forschungsgemeinschaft, Bonn Bad-Godesberg (Germany).

REFERENCES

Gotoh, S. "Reduction Processes in Waterlogged Soils with Special Reference to Transformation of Nitrate, Manganese and Iron," Ph. D. Thesis, Kyushu University, Japan (1974).

Gotoh, S. and W. H. Patrick. "Transformation of Iron in a Waterlogged Soil as Influenced by Redox Potential and pH," *Soil Sci. Soc. Am. Proc.* 38:66-71 (1974).

Hammann, R., J. C. G. Ottow. "Reductive Dissolution of Fe_2O_3 by Saccharolytic Clostridia and *Bacillus polymyxa* under Anaerobic Conditions," *Z. Pflanzenern Bodenk*. 137: 108-115 (1974).

Jacob, H. E. "Redoxpotentials," in *Meth. in Microbiol.* Vol. 2 (New York: Academic Press, 1970), pp. 92-121.

Mokrasch, L. C. "Analysis of Hexosephosphates and Sugar Mixtures with Anthrone Reagent," *J. Biol. Chem.* 208:55-59 (1954).

Munch, J. C. and J. C. G. Ottow. "Modelluntersuchungen zum Mechanismus der Bakteriellen Eisen-Reduction in Hydromorphen Böden," *Z. Pflanzenern. Bodenk.* (in press).

Ottow, J. C. G. "Selection, Characterization and Iron-Reductaseless (nit⁻) Mutants of Iron-Reducing Bacteria," *Z. Allg. Mikrobiol.* 10:55-62 (1970).

Ottow, J. C. G. and H. Ottow. "Gibt es eine Korrelation Zwischen der Eisenreduzierenden und Nitratreduzierenden Flora des Bodens?" *Zentbl. Bakt. Abt. II* 124:314-318 (1970).

Ottow, J. C. G. "Iron Reduction and Gley Formation by Nitrogen-Fixing Clostridia," *Oecologia* 6:164-175 (1971).

Ottow, J. C. G. and H. Glathe. "Pedochemie und Pedomikro-biologie Hydromorpher Böden. Merkmale, Voraussetzungen und Ursache der Eisenreduktion," *Chemie Erde* 32:1-43 (1973).

Patrick, W. H. and F. T. Turner. "Effect of Redox Potential on Manganese Transformation in Waterlogged Soil," *Nature* 220:476-478 (1968).

Ponnamperuma, F. N., W. L. Yuan and M. T. M. Nhung. "Manganese Dioxide as a Remedy for a Physiological Disease of Rice Associated with Reduction of Soil," *Nature* 207: 1103-1104 (1965).

Ponnamperuma, F. N. "The Chemistry of Submerged Soils," *Adv. Agron.* 24:29-96 (1972).

Schlichting, E. and H. Blume. *Bodenkundliches Praktikum* (Hamburg and Berlin: Parey Verlag, 1966).

Schnitzer, M. and S. I. M. Skinner. "Organometallic Inter-actions in Soils," *Soil Sci.* 4:247-252 (1967).

Szilagyi, M.   "Reduction of Fe^{3+} Ion by Humic Acid Preparations," *Soil Sci.* 111:233-235 (1971).

Takai, Y. and T. Kamura.   "The Mechanism of Reduction in Waterlogged Paddy Soil," *Folia Microbiologia* 11:304-313 (1966).

Yoshida, T.   "Microbial Metabolism of Flooded Soils," in *Soil Biochemistry*, Vol. 3, E. A. Paul and A. D. McLaren, eds. (New York: Marcel Dekker, Inc., 1975), pp. 83-112.

Yuan, W. L. and F. N. Ponnamperuma.   "Chemical Retardation of the Reduction of Flooded Soils and the Growth of Rice," *Plant Soil* 25:347-360 (1966).

CHAPTER 38

MOBILIZATION, TRANSPORT AND MICROBIALLY ASSISTED PRECIPITATION OF IRON IN A BOG CREEK

P. HALBACH
K. H. UJMA

Research Group "Sedimentary Ore Formation"
Mineral.-Petrograph. Institut
Techn. Univ. Clausthal
D-3392 Clausthal-Zellerfeld, Federal Republic of Germany

INTRODUCTION

In some regions of the Harz mountains ombrogenous raised bog areas are prevalent. The investigated location is situated west of the Oderteich, a reservoir about 6 km north of the ancient mining town of St. Andreasberg. The area lies about 730 m above mean sea level. It is drained by a few bog creeks that flow into Lake Oderteich on its western bank. One of these bog creeks, showing voluminous iron hydroxide precipitates, runs into the reservoir about 500 m north of the dam. This creek was the main subject of our studies.

The climate in this environment can be described generally as perhumid-subnival. The average air temperature is 3°C; the amount of precipitation is 1500 to 1600 mm per year. In warmer times of the former Holocene the climate was more favorable for the bog growth. The deepest peat layers have been formed in the Atlanticum (interglacial stage of Holocene). The sedentary Sphagnum peat covering has a recent maximum thickness of 1.5 m; its humosity is relatively low.

The bedrock in the surroundings of Lake Oderteich is part of the micropegmatitic granite of the western margin of the Brocken batholith (Figure 1). We consider this granitic bedrock to be the source of metal supply of the above mentioned creek. Towards the west this granite is in contact with the older rocks of the Ackerbruchberg-quarzite. The upper course of the creek is situated in this contact zone.

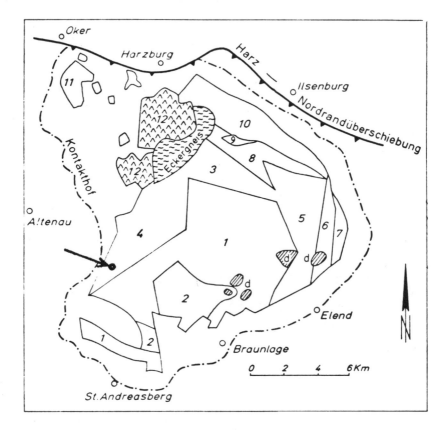

Figure 1. Rock varieties of the Brocken granite (according
to Mohr, 1973). *Legend*: (1) roof granite (2) porphyrite
roof granite (3) micropegmatitic granite of the Hercynian
northern marginal zone (4) micropegmatitic granite of the
western margin (5) micropegmatitic granite of the eastern
margin (7) eastern marginal diorite (8) Hercynian granite-
diorite zone (9) northern marginal diorite (10) Ilsestein-
granite (11) Oker-granite (12) Harzburg-gabbro. The arrow
marks the position of the investigated area.

RESULTS

The granite rock samples have a porphyric texture with phenocrysts of quartz and albite. The groundmass shows orthoclases and quartz with micrographic intergrowth (Figure 2). A further component is <u>biotite</u>; tourmaline and oxides of iron

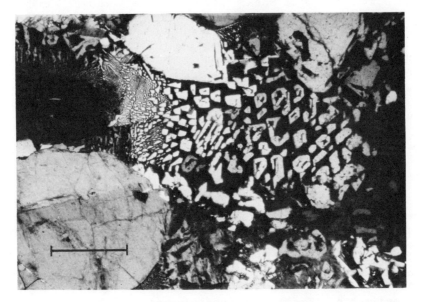

Figure 2. Microphotograph of granite: Micrographic intergrowth of quartz and orthoclase. Thin section, crossed nicols, scale bar: 1 mm.

and titanium are accessories. Microscopic analyses show contents of 35–40 vol % quartz, 45–50 vol % orthoclase, 5–15 vol % plagioclase, and 5–10 vol % biotite and accessories. Table I shows a comparison of results of chemical analyses

Table I
Chemical Analyses of Granite, Samples

Sample	SiO_2 (%)[a]	Al_2O_3 (%)	K_2O (%)	Na_2O (%)	Fe (%)	CaO (%)	MgO (%)	TiO_2 (%)	P (%)	Mn (%)
Nr. 4[b]	74.79	12.90	5.16	3.1	1.42	0.76	0.23	0.23	240	420
Nr. 5[b]	74.14	13.20	5.19	3.01	1.52	1.15	0.23	0.28	280	410
GR 4	73.34	14.40	4.8	2.92	0.91	0.46	0.25	0.22	240	280
GR 5	73.66	14.16	6.12	2.2	0.88	0.26	0.35	0.21	200	200

[a]All values: wt % of dry substance.
[b]According to Seim and Eidam (1974).

of two partially weathered granite samples to values obtained
from Seim and Eidam (1974). The comparison reveals no significant changes of SiO_2 content, but an enrichment of Al_2O_3
and a loss of Fe and Ca. The expected losses of alkali
metals are not significant.

About 200 m above the creek mouth a lateral spring
rises out of the peat bank. Its water has a remarkably high
content of dissolved iron, *i.e.*, 10-15 mg/l. Rather constantly
throughout the year the spring produces 6-7 liters of water
per minute with a temperature of about 5°C; this is ortho-
thermal behavior. A calculation shows that this spring produces about 50 kg of iron per year. Taking into account a
catchment drainage area of 0.03 km² and a weathering rate of
only 0.1 mm per year of the granite surface, that quantity of
iron may be released. In this calculation we assumed that
only 40 percent of the iron content of the granite rock was
dissolved in water during the weathering process. The iron
originates from mineral phases as hematite, goethite, or
mafic components as biotite.

Immediately below the spring orifice an aqueous
deposit, the so-called sinter terrace, has been formed. It
consists mainly of yellowish-red "ferric hydroxide". These
substances contain up to 48 percent Fe (Table II). Furthermore the content of P and organic C is remarkable. The
enrichment of trace metals is sometimes considerable; the
highest observed concentration of Zn was 1700 ppm. The bottom
of the creek, too, is covered with reddish-brown iron oxyhydrate sediments. The iron mineral phases of all sediment
samples proved to be mainly amorphous. By aging and dehydration an increase of crystallinity takes place. Later on
metastable ferrihydrite (Chukrov, 1974) and finally goethite
are found as phases.

We investigated creek and spring water by means of
physicochemical and chemical methods in the field and in the
laboratory. The data in Table III are typical. Summarizing
these data, the creek water is a normal air-saturated dystrophic
surface water, whereas the spring water shows features of a
microaerobic bicarbonate water.

Table IV shows a profile of spring and creek water data
in comparison with the computed Fe(II) concentration in
equilibrium for the measured pH and redox-potential values
according to the Nernst-equation:

$$Eh = E_o + \frac{R \cdot T}{n \cdot F} \cdot \ln \frac{a_{Fe(OH)_3} \cdot a_{H^+}^3}{a_{Fe^{++}} \cdot a_{H_2O}^3}$$

$$= 1293 - 177 \, pH \quad mV$$

Table II

Chemical Analyses of Terrace Deposits

Sample	Fe (%)[a]	Mn (%)	P_2O_5 (%)	S (%)	SiO_2 (%)	Al_2O_3 (%)	CaO (%)	MgO (%)	C[b] (%)	Na_2O (%)	K_2O (%)	Zn (%)
B	44.00	0.009	1.70	0.13	2.80	0.21	0.10	0.11	0.54	<0.01	<0.01	0.010
D	40.60	0.002	2.68	0.18	2.36	0.49	0.10	0.10	2.97	0.01	<0.01	0.010
E	46.60	<0.001	1.67	0.52	2.36	0.88	0.10	0.09	0.46	0.01	<0.01	0.003
G	48.00	n.d.	n.d.	n.d.	3.40	0.17	0.10	0.09	n.d.	0.01	<0.01	0.170

[a]All contents: wt % of dry substance.
[b]Organic carbon only; CO_2 not detectable.

Table III
Analysis of Spring and Creek Water

	Spring Water	Creek Water
pH	5.8	6.0
Eh	185 mV	320 mV
Fe^{++}	10 to 15 mg/l	1 to 6 mg/l
HCO_3^-	± 34 mg/l	± 14 mg/l
Diss. O_2	0.2 ppm	± 8 ppm
Color	20 ppm Pt	± 40 ppm Pt

Table IV
Profile of Spring and Creek Water Data
and Computed Fe(II) Concentrations

Sample Test Point	pH	Eh^a (V)	Analyzed Fe(II) Concentration $(10^{-6}$ m/l)	Max. Theoretical Fe(II) Concentration $(10^{-6}$ m/l)
P 2	5.85	0.184	292	1.760
P 1	5.72	0.172	276	6.910
P 3	5.76	0.173	274	5.050
P 5 a	5.74	0.200	274	2.020
P 5	5.80	0.196	272	1.560
P 7	5.77	0.285	122	59.5
P 8	5.80	0.319	68.0	12.8
P 9	5.99	0.312	104	4.54
P 10	5.77	0.343	59.1	6.19
P 11	6.05	0.306	82.4	3.79
P 12	6.27	0.304	66.3	0.897

[a] Corrected for standard conditions.

        In the upper part of Table IV (samples P 2 to P 5)
gives values for terrace water. The double line is the boundary
between them and the creek water (P 7 to P 12). The transition
from terrace to creek water clearly shows that on the terrace
the properties of the water are still determined by the ground-
water character of the spring. Here the water seems to be
an undersaturated solution of Fe(II). On the other hand,
there is a calculated supersaturation of Fe(II) in the creek
water.

        The iron-containing deposits consist of aqueous sedi-
ments of biological and chemical origin. The proportion of
biological and purely inorganic precipitation is a function
of the physicochemical conditions that moreover fix the fields
of existence of microorganisms. The terrace is dominated by
the microaerobic, psychrophilic *Gallionella ferruginea*
(Figure 3). Static culture experiments were made with natural

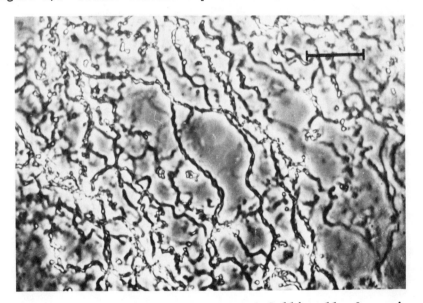

Figure 3. Microphotograph of stems of *Gallionella ferruginea*.
   Phase contrast, scale bar: 10 m.

water samples. Half of them were sterilized to test inorganic
iron precipitation in comparison with the process under the
influence of microbiological activity. All samples were
maintained under microaerobic conditions. The results showed
that the sterile samples had not yet begun to lose iron by
precipitation when the Gallionella-active samples had precipitated all iron as oxyhydrate with no detectable remaining
iron in solution.

DISCUSSION

As cited above, the iron is found primarily in the granite that represents the bedrock for the sedentary Sphagnum peat layers. Usually the mafic components of magmatic rocks are rather unstable to chemical weathering in humid climates, which means that under suitable thermodynamic conditions iron is mobilized from biotite as well as from iron oxide components of the granite. Dark micas, *e.g.*, biotite, decompose to clay minerals; pH controls mainly the speed and the product of disintegration. Due to special properties of the humic substances and independence of the primary mobilization of iron from the granite, a buffering accumulation of Fe(II) or various Fe(II)-aquo-hydroxo ions (Schnitzer and Khan, 1972) in the humic matter of the peat is probable.

During the process of peat formation from plant tissue, reducing and acid conditions predominate. Therefore, Fe(III) can be reduced and dissolved in the contact of peat and granite. In the spring water, divalent iron is mobilized as hydrogen carbonate and transported as ions in aqueous solution (Halbach, 1975). The organic complex compounds of iron with fulvic and humic acids are more important for the transport of iron under the conditions of surface waters with high oxygen content, *i.e.*, increased redox potential, than under the above-described conditions of mobilization. It seems that in surface bog waters with high oxygen content humic complexes of Fe(II) retard an immediate alignment of the redox equilibrium Fe(II) /Fe(III), as the increased values for Fe(II) show. Besides, a certain amount of colloidal Fe(III)-oxyhydrate particles must be supposed. Their size is approximately several 10 nanometers, making them easily transportable as suspended load.

The fact that in spite of undersaturation a precipitation of iron-containing compounds takes place on the terrace shows that these precipitations can only be explained as a microbial process. Looking at Figure 4, point A, iron oxidation following the equation

$$Fe^{++} + 3\ H_2O = Fe(OH)_3 + 3\ H^+ + e^-$$

should not take place there. (This equation is certainly a simplifying description of the complicated natural process, but we assume that it gives a good trend.) Obviously the iron oxidizing bacterium *Gallionella ferruginea* is able to use the dissolved Fe(II) as autotrophic organism and to precipitate it, regardless of the solubility product of ferric hydroxide and the external chemical parameters of the water (Ujma, 1976). This hypothesis is confirmed by out static culture precipitation experiments where the sterilized samples behaved in a completely different manner than those with natural microbial activity. Fe(III), which is bound to the organic matter of

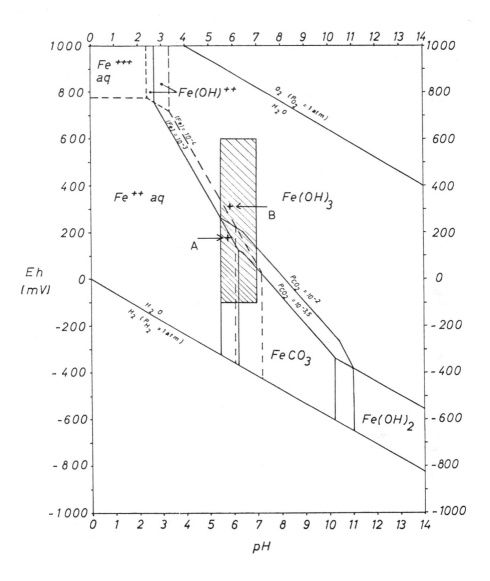

Figure 4. Eh–pH diagram with stability field of "ferric hydroxide " The hatched area marks values for some natural environments.

stems of *Gallionella*, can be visualized by stain reactions resulting in Berlin blue. On this organic complex compound Fe(III) is protected from mineralization to ferric hydroxide for a short time, until the aging of the stems has progressed and the formation of microscopically well-perceptible Fe(III)-hydroxide particles begins (Hanert, 1973). The further accumulation of these Fe(OH)$_3$-particles seems to be a surface microprocess rather than a microbial one. The precipitation of iron in the creek will mainly take place chemically (Figure 4, point B). Their fine grain size enables organic colloidal particles to deposit as iron humates only in an environment with a very low velocity of flow, *i.e.*, in Lake Oderteich.

ACKNOWLEDGMENTS

We wish to thank our colleague H. Hanert, TU Braunschweig, who supported us with valuable assistance in microbiology, and the company Stahlwerke Peine-Salzgitter A.G. who helped us in doing several chemical analyses. The whole study was kindly sponsored by a grant of Deutsche Forschungsgemeinschaft.

REFERENCES

Chukrov, F. W. "Uber die Natur der Eisenoxide in geologisch jungen Bildungen," *Chemie der Erde* 33:109-124 (1974).

Halba h, P. "Mineralogical and Geochemical Investigations on Finnish Lake Ores," *Bull. Geol. Soc. Finland* 48:33-42 (1975).

Hanert, H. "Quantifizierung der Massenentwicklung des Eisenbakteriums *Gallionella ferruginea* unter natürlichen Bedingungen," *Archiv für Mikrobiologie* 88:225-243 (1973).

Mohr, K. *Harz, westlicher Teil; Geologischer Fuhrer Nr. 58* (Berlin: Bornträger, 1973.)

Schnitzer, M. and S. U. Khan. *Humic Substances in the Environment* (New York: Marcel Dekker, Inc., 1972).

Seim, R. and J. Eidam. "Vergleich chemischer Untersuchungen der Granite des Brocken- und Rambergmassivs im Harz," *Chemie der Erde* 33:31-46 (1974).

Ujma, K. H. *Geochemische Untersuchungen zum Löslichkeits- und Ausfällungsverhalten von Eisen in rezenten Fe(II)- haltigen Wassern des Hochmoorgebietes "Drei Horste" am Oderteich (Harz)* Diplomarbeit, Techn. Univ. Clausthal (1976).

CHAPTER 39

THE PRECIPITATION OF IRON AND MANGANESE IN FENNOSCANDIA: GEOLOGY AND GEOCHEMISTRY

LIISA CARLSON
TAPIO KOLJONEN
ANTTI VUORINEN

Department of Geology and Mineralogy
University of Helsinki
P.O. Box 115
SF-00171 Helsinki 17, Finland

INTRODUCTION

Most of the iron and manganese precipitates of Fennoscandia are lake and bog ores collected from southern areas (Vogt, 1915; Aarnio, 1915; Ljunggren, 1953; Alfsen and Christie, 1972; Halbach, 1976). Ljunggren (1953) described manganese dioxide crusts on stones in a river in southern Sweden. Recently, various types of iron and manganese precipitates in Finland have been studied by Vasari *et al.* (1972), Alhonen *et al.* (1975), and Koljonen *et al.* (1976).

This chapter will discuss the principles of precipitation of iron and manganese, and the distribution of various types of precipitates in Fennoscandia. According to Aarnio (1915), most Finnish lakes bearing lake ore are situated in the central and eastern parts of the country, between N Lat. 61 and 65°, the distribution being dependent on both climate and the type of sediments covering the bedrock. For this study various types of iron and manganese precipitates were collected from different parts of Finland and northern Norway. The sampling locations are presented on a map of Quaternary deposits in Finland (Figure 1).

CHEMICAL COMPOSITION

For chemical analysis the precipitates were concentrated (Koljonen *et al.*, 1976, p. 122) and dissolved in HNO_3 and H_2O_2. Fe was determined by titration, Mn colorimetrically, and Na, K, Mg, Ca, Co, Ni, Zn, Cu, Cr and Pb as well as minor concentrations of Fe and Mn by AAS.

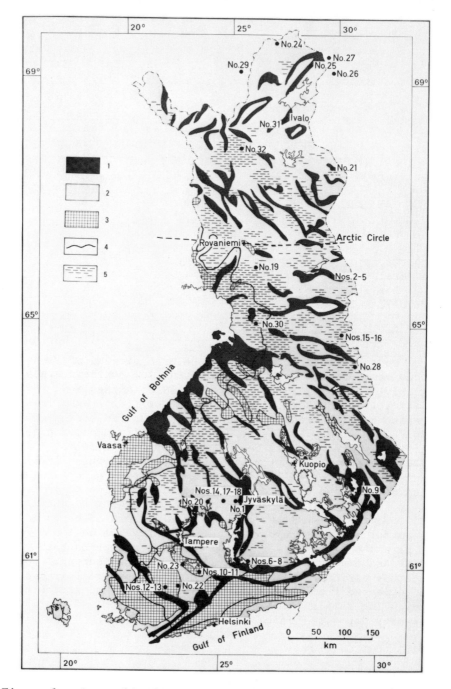

Figure 1.   Generalized map of Quaternary deposits in Finland
   (Topografikunta and Suomen maantieteellinen seura 1968) with
   sampling locations (Nos. 1–32).   (1) Sand and gravel.   The
   areas are somewhat exaggerated;   (2) Till;   (3) Clay and
   silt;   (4) The highest shoreline of the Littorina Sea;
   (5) Peat bogs.

The samples, grouped according to the mode of occurrence, are described in Tables I and II. The chemical composition of the precipitates is presented in Tables III and IV. The mineral composition of iron and manganese precipitates in glaciofluvial material is discussed in Koljonen *et al.* (1976) and in Carlson *et al.* (1977).

DISCUSSION

In the lowlands of Fennoscandia peat bogs are widespread and podzol the most common soil type (*cf.*, Dudal *et al.*, 1966). In both of them the decomposition of plant material leads to the lowering of Eh and pH of water, which consequently leaches iron and manganese out of the mineral material it flows through. Part of the iron is precipitated in the B horizon of podzol where Eh and pH are seldom high enough for manganese to be precipitated. Manganese and the rest of the iron migrate mainly as organic complexes until they reach areas where Eh and pH conditions favor their precipitation. The Mn/Fe ratio of groundwater is *ca.* 0.04 (Koljonen *et al.*, 1976).

The most common types of iron and manganese precipitates are those found in glaciofluvial material and lake and bog ores. When ground water with dissolved Fe and Mn complexes flows through permeable glaciofluvial material, its Eh and pH gradually increase through hydrolysis, and the complexes are in part broken down. The iron set free may oxidize and precipitate, but the Eh and pH conditions are not high enough for the manganese to be oxidized and it migrates as Mn^{2+} ion. When ground water reaches aerated areas, *e.g.*, in coarse-grained layers or at the top of the ground water table, the manganese is oxidized and precipitated by atmospheric oxygen. The iron is oxidized thereafter when the more stable iron complexes have been broken down. That manganese has been precipitated before iron is clear, for example, from concentric spring deposits (Carlson *et al.*, 1977). When manganese is precipitated from solutions containing both elements, considerable amounts of iron are coprecipitated, but the Mn/Fe ration of iron precipitates is extremely low (Tables III and IV, Nos. 1-14 and 24-27, and Koljonen *et al.*, 1976).

The precipitation of iron and manganese in glaciofluvial material is supposed to be inorganic. Manganese in particular is rapidly oxidized and precipitated. Moreover, the milieu where the precipitates are formed does not favor any bacterial activity. The various types of precipitates described by Koljonen *et al.*, (1976) are formed in all parts of Fennoscandia where eskers, deltas and other glaciofluvial deposits containing coarse-grained material are found. In northern Fennoscandia the iron precipitates are dark brown

Table I

Description of Samples, Nos. 1-23

Number	Location[a]	N Lat.	E Long.	Description
Precipitates in Glaciofluvial Material				
1	Jyvaskyla, Palokangas	62.2°	25.7°	Hard black cement in gravel
2	Kuusamo, Taviharju	65.9°	28.9°	Black cement in gravelly sand
3	Kuusamo, Taviharju	65.9°	28.9°	Black cement in fine sand
4	Kuusamo, Taviharju	65.9°	28.9°	Black cement in sand
5	Kuusamo, Taviharju	65.9°	28.9°	Black cement in gravelly sand
6	Heinola	61.3°	26.0°	Black precipitate in a spring in gravel
7	Heinola	61.3°	26.0°	Brown precipitate in a spring in gravel
8	Heinola	61.3°	26.0°	Loose black precipitate in gravel, a few years old
9	Pyhaselka, Hammaslahti	62.4°	30.0°	Black cement in gravel
10	Hameenlinna, Hattelmala	61.0°	24.4°	Black precipitate in stony gravel
11	Hameenlinna, Hattelmala	61.0°	24.4°	Brown precipitate in sandy gravel
12	Somero, Hantala	60.4°	23.3°	Black precipitate in a spring in gravelly sand
13	Somero, Hantala	60.4°	23.3°	Brown precipitate in a spring in gravelly sand
14	Petajavesi	62.2°	25.1°	Brown precipitate in a spring
Lake Ores				
15	Suomussalmi, Lake Aittojarvi	65.0°	29.5°	Crust ore, depth < 0.7 m
16	Suomussalmi, Lake Aittojarvi	65.0°	29.5°	Penny ore, depth *ca* 0.5 m
17	Petajavesi, Lake Kirrinjarvi	62.2°	25.1°	Penny ore, depth 1.2-1.5 m
18	Petajavesi, Lake Kirrinjarvi	62.2°	25.1°	Pisolithic ore, depth 1.5-1.7
Miscellaneous				
19	Ranua, Lake Saukkojarvi	66.2°	26.2°	Glass-like brown cement in sand, small pieces on lake bottom, depth 0.3-0.7 m
20	Keuruu, Huttula	62.2°	24.7°	Black crust on a stone in a rivulet
21	Savukoski, Sokli	67.8°	29.4°	Fluvially transported and redeposited preglacial weathering crust
22	Tammela, Torronsuo	60.7°	23.6°	Hard brown cement in gravel, close to a peat bog
23	Kylmakoski, Lake Jalanti	61.1°	23.8°	Glass-like brown cement in till, an ancient shoreline

[a]Finland, unless otherwise stated.

Table II

Description of Samples, Nos. 24-32

Number	Location	N Lat.	E Long.	Description
Precipitates in Glaciofluvial Material				
24	Utsjoki	69.8°	27.0°	Brown cement in gravelly sand
25	Inari, Sevettijarvi	69.6°	28.8°	Dark brown cement in stony gravel
26	Norway, Svanvik	69.4°	30.0°	Dark brown cement in sandy gravel
27	Norway, Neiden	69.7°	29.4°	Dark brown hard cement in sandy gravel
Lake Ores				
28	Kuhmo, Lake Iso Kuhmojarvi	64.6°	29.7°	Black precipitate resembling crust ore, depth 0.6-1.0 m
29	Norway, Karasjok	69.5°	25.6°	A layer of black cemnt overlying a brown one, both *ca.* 2 mm thick, depth < 1 m
Miscellaneous				
30	Ulikiiminki, Palokangas	65.2°	26.3°	Dark brown cement in sand in the middle of a bog
31	Inari, Ivalo	68.7°	27.5°	Black crust on sand particles in a fluvial deposit
32	Kittila, Pailakainen	68.1°	25.6°	Dark brown cement in gravelly sand, redeposited preglacial weathering crust

[a]Finland, unless otherwise stated.

Table III

Chemical Composition and Mn/Fe Ratio of Precipitates, Nos. 1-23

Number	Fe(%)	Mn(%)	Mn/Fe	Na(%)	K(%)	Mg(%)	Ca(%)	Co(ppm)	Ni(ppm)	Zn(ppm)	Cu(ppm)	Cr(ppm)	Pb(ppm)
Precipitates in Glaciofluvial Material													
1	2.3	57.6	25.0	0.18	0.18	0.34	0.07	42	40	92	45	40	110
2	5.6	47.3	8.5	0.01	0.51	0.28	<0.01	<80	n.d.	60	n.d.	n.d.	n.d.
3	10.8	42.5	3.9	0.10	0.69	0.30	0.01	<70	n.d.	90	n.d.	n.d.	n.d.
4	10.8	34.7	3.2	1.33	0.68	0.50	<0.02	<130	n.d.	690	n.d.	n.d.	n.d.
5	7.0	0.11	0.02	1.21	0.71	0.59	<0.01	<90	n.d.	370	n.d.	n.d.	n.d.
6	6.0	44.7	7.5	0.00	0.63	0.23	0.03	853	405	1106	495	45	90
7	52.4	2.4	0.05	0.01	0.38	0.66	0.02	434	130	374	430	394	160
8	10.5	38.4	3.7	1.33	1.21	0.93	0.01	411	1805	620	968	118	120
9	5.2	44.3	8.5	0.01	0.01	1.53	0.10	319	900	86	198	38	65
10	11.6	43.1	3.7	0.00	0.80	0.46	0.69	127	960	1510	890	46	90
11	51.2	1.1	0.02	1.20	2.80	1.70	0.19	171	45	372	35	41	55
12	19.9	38.4	1.9	0.77	0.77	0.61	0.38	65	600	121	111	31	80
13	61.9	0.60	0.01	0.12	0.26	0.24	0.06	669	50	449	319	103	95
14	49.2	0.34	0.01	0.37	0.00	0.20	0.53	93	17	14	6	29	45
15	23.4	27.9	1.2	0.03	0.21	0.07	0.16	236	34	720	7	21	60
16	3.5	25.3	0.8	0.06	0.21	0.09	0.25	261	60	640	8	24	65
17	52.4	11.7	0.22	0.16	0.14	0.05	0.06	371	50	474	10	44	80
18	55.6	14.5	0.26	0.06	0.13	0.03	0.10	78	35	390	10	40	70
19	31.3	1.0	0.03	0.01	0.01	0.01	0.11	47	20	164	4	19	40
20	23.6	29.9	1.3	0.03	0.16	0.01	0.11	518	30	300	5	33	140
21	37.3	12.8	0.34	0.00	0.02	0.37	1.02	76	33	6175	140	30	145
22	46.0	0.6	0.01	0.06	0.30	0.27	0.08	190	90	151	180	170	135
23	24.1	0.05	0.00	0.32	0.33	0.52	0.02	48	30	146	24	80	70

n.d. = not determined

Table IV

Chemical Composition and Mn/Fe Ratio of Precipitates, Nos. 24–32

Number	Fe(%)	Mn(%)	Mn/Fe	Co(ppm)	Ni(ppm)	Zn(ppm)	Cu(ppm)	Cr(ppm)	Pb(ppm)
Precipitates in Glaciofluvial Material									
24	11.9	0.25	0.02	54	30	58	90	141	<10
25	11.9	0.13	0.01	35	20	46	97	148	100
26	8.5	0.57	0.07	55	30	60	97	168	130
27	11.6	0.69	0.06	147	40	62	170	257	90
Lake Ores									
28	8.6	33.9	3.9	488	410	1651	15	36	60
29	36.9	16.6	0.4	112	330	355	35	85	33
Miscellaneous									
30	2.3	0.01	0.00	6	10	28	10	84	60
31	6.4	53.7	8.4	58	100	190	105	63	120
32	7.5	0.02	0.00	47	40	40	35	177	<10

because large amounts of highly polymerized organic compounds
are coprecipitated. Consequently the concentration of Fe and
Mn shown in Table IV are generally low.

Lake ores are formed of iron and manganese brought to
the bottom sediments by ground water and by slow deposition
of fluvially transported colloidal suspensions and organic
complexes (Halbach, 1976). In southern and central Finland
the lakes are rich in organic matter, the decomposition of
which consumes oxygen and causes reducing conditions in the
bottom mud. Iron and manganese are set free and diffuse as
Fe^{2+} and Mn^2 ions to shallow parts of the lake where oxidizing
conditions prevail at the sediment—water interface (*cf.*,
Strakhov, 1966). Lake ore is formed mainly as a belt close
to the shore. Pisolithic ores (< 1 cm) are formed in
slightly deeper parts where Eh is a bit lower than where penny
(1 – 3 cm) and crust ores are formed. The Mn/Fe ratio is
thus lowest in pisolithic and highest in crust ores. In
southern Fennoscandia the temperature of lake water varies
within wide limits in the course of a year and little organic
matter is deposited during winter. The Eh and pH conditions
fluctuate and therefore the belt of precipitation is broad
and the lake ores consist of alternating Fe- and Mn-rich
layers.

In northern Finland and adjoining areas, as well as in
the deepest parts of large lakes, less organic matter is
deposited and decomposed. The temperature of the water in
such lakes remains low even in summer, and Eh and pH fluctuate
much less. Lake ores are formed as a narrow and vertically
thin belt without clear division into various ore types.
Alternating layers are seldom formed. It seems that the Mn/Fe
ratio in lake ores generally increases towards the north.
Variation in manganese content is great even within one lake,
but more iron is precipitated in glaciofluvial material in
northern than in southern Fennoscandia, and in lakes poor in
organic matter Fe-rich precipitates are formed as layers in
bottom sediments because of the mainly vertical diffusion
(*cf.*, Strakhov, 1966). Lake ores are uncommon and very small
north of the Arctic Circle because most of the iron and manganese is precipitated before reaching the lake. In Sodankyla
(N Lat. 67.5°) precipitates resembling lake ore have been
found in a brook flowing through a peat bog. The northernmost
lake ore ever described is that from Karasjok, Norway (Tables
II and IV, No. 29).

According to Varentsov (1972), lake ores can be formed
by inorganic precipitation, by "very selective, chemosorption
accumulation with autocatalytic effect." Manganese-precipi-
tating bacteria have been identified in Finnish lake ore
(Halbach, 1976), but it is unlikely that microorganisms are
essentially involved in the formation of the ores.

In peat bogs iron is precipitated as hydroxide, carbonate or phosphate. Typical of the milieu is low pH and high content of CO_2 and dissolved organic compounds in the water. In surface layers where oxidizing conditions prevail, iron is oxidized and precipitated as hydroxide. The precipitation is aided by evaporation. Between peat layers and on the bottom of bogs where reducing conditions prevail, iron is precipitated as siderite ($FeCO_3$), and where phosphate concentration is high enough, as vivianite, $FeHPO_4$ (Puustjarvi, 1952). Thin (< 1 cm) vivianite layers are also found in reducing bottom sediments of some lakes. No manganese is precipitated in bogs (Table IV, No. 30) because high enough Eh and pH are not reached. Iron precipitates in bogs are common in central and northern Finland, especially around the northern part of the Gulf of Bothnia.

Black crusts are frequently formed on stones in rivers (Tables I and III, No. 20). In an early study, Ljunggren (1953) reported the precipitation of manganese and iron catalyzed by bryophytes. However, inorganic precipitation seems to be more common and to occur where Eh and pH are locally raised. Iron is precipitated first, catalyzed by stone surfaces, and concentric layering resembling that of lake ores occurs. Crusts on stones are also formed on lake shores where wave action is significant. Manganese is precipitated above the water level and iron somewhat below it because of the difference in Eh. No crust on stones is formed on the shore of the Arctic Ocean in northern Norway.

Manganese is separated from iron in sedimentary processes because the stability fields of possible minerals lie at slightly different values of Eh and pH (Krauskopf, 1957) and because of differences in complex stability. Microorganisms may aid the selective oxidation and precipitation, which nevertheless is mostly inorganic. In glaciofluvial deposits iron and manganese hydroxide cements occur as separate belts. The youngest Mn-rich precipitates may contain appreciable amounts of iron, up to an Mn/Fe ratio of *ca.* 2, but further separation occurs in the wet precipitate by diffusion, and the hard precipitates are poor in Fe. The Mn-content of iron precipitates is low, as a rule (Table III).

Iron and manganese precipitates are effective adsorbers of ions from solution (Goldberg, 1954). In terrestrial precipitations, the association of manganese with cobalt has been especially observed (Taylor and McKenzie, 1966). In this study, Ni more clearly than Co and Zn seems to follow Mn (Tables III and IV). Local variation occurs, reflecting the trace element concentration in water. In lake ores the Co and Pb content is about the same as in nodules from the Baltic (Winterhalter, 1966), the content of Zn slightly higher and of Ni, Cu and Cr lower. In deep-sea nodules more trace elements are adsorbed because the rate of deposition and nodule formation is slower.

REFERENCES

Aarnio, B. "Uber die Ausfällung des Eisenoxyds und der Tonerde in Finnlandischen Sand- und Grusboden," *Geotek. Tied.* 16 (1915).

Alfsen, B. E. and O. H. J. Christie. "Analyses of a Sedimentary Iron Ore Pisolith from Lake Storsjoen, South Norway," *Nature (Phys. Sci.)* 237:125-126 (1972).

Alhonen, P., T. Koljonen, P. Lahermo and R. Uusinoka. "Ferruginous Concretions Around Root Channels in Clay and Fine Sand Deposits," *Bull. Geol. Soc. Finland* 47:175-181 (1975).

Carlson, L., P. Lahermo and T. Koljonen. "Mineralogy and Chemistry of an Oxidate Precipitate in a Spring in Somero, SW Finland," *Third Meeting of the European Clay Groups, Oslo,* (1977).

Dudal, R., R. Tavernier and D. Osmond. "Soil Map of Europe 1:2.500.000, Explanatory Text," *FAO* Rome (1966).

Goldberg, E. D. "Marine Geochemistry. 1. Chemical Scavengers of the Sea," *J. Geol.* 62:249-265 (1954).

Halbach, P. "Mineralogical and Geochemical Investigations on Finnish Lake Ores," *Bull. Geol. Soc. Finland* 48:33-42 (1976).

Koljonen, T., P. Lahermo and L. Carlson. "Origin, Mineralogy, and Chemistry of Manganiferous and Ferruginous Precipitates Found in Sand and Gravel Deposits in Finland," *Bull. Geol. Soc. Finland* 48:111-135 (1976).

Krauskopf, K. B. "Separation of Manganese from Iron in Sedimentary Processes," *Geochim. Cosmochim. Acta* 12:61-84 (1957).

Ljunggren, P. "Some Data Concerning the Formation of Manganiferous and Ferriferous Bog Ores," *Geol. Foren. Stockh. Forh.* 75:277-297 (1953).

Puustjarvi, V. "The Precipitation of Iron in Peat Soils," *Acta Agral. Fennica* 78:1 (1952).

Strakhov, N. M. "Types of Manganese Accumulation in Present-day Basins: Their Significance in Understanding of Manganese Mineralization," *Int. Geol. Rev.* 8:1172-1196 (1966).

Taylor, R. M. and R. M. McKenzie. "The Association of Trace
Elements with Manganese Minerals in Australian Soils,"
Aust. J. Soil Res. 4:29-39 (1966).

Varentsov, I. M. "Geochemical Studies on the Formation of
Iron-Manganese Nodules and Crusts in Recent Basins,"I.
Eningi-Lampi Lake, Central Karelia," *Acta Miner. Petrogr.*
Szeged, 20:363-381 (1972).

Vasari, Y., T. Koljonen and K. Laakso. "A Case of Manganese
Precipitate in the Taviharju Esker, Kuusamo, North East
Finland," *Bull. Geol. Soc. Finland* 44:133-140 (1972).

Vogt, J. H. L. "Om manganrik sjømalm i Storsjøen, Nordre
Odalen," *Nor. Geol. Unders.* 75:6 (1915).

Winterhalter, B. *Iron-manganese Concretions from the Gulf of
Bothnia and the Gulf of Finland.* English summary, Geotek.
Julk. 69 (1966).

CHAPTER 40

THE DISTRIBUTION OF SULFUR
IN ORGANIC FRACTIONS OF PEAT:
ORIGINS OF SULFUR IN COAL

D. J. CASAGRANDE
K. GRONLI
V. PATEL

College of Environmental and Applied Sciences
Governors State University
Park Forest South, Illinois   60466   USA

INTRODUCTION

The utilization of high sulfur coal has numerous
environmental problems. While the gross forms of sulfur
(pyrite, sulfate, organic sulfur) have been identified and
reported (Ruch *et al.*, 1974; and many others), the characterization of the organic sulfur in coal along with a knowledge
of the geographic distribution of sulfur forms in a coal
seam will allow for (1) the development of technology to
remove organic sulfur in coal and (2) the selective mining
of areas of a coal basin with lower levels of sulfur.

The approach of the present investigation is to study
the progenitors of coal in the hope of learning about the
processes by which the various forms of sulfur are incorporated
and distributed in potential coal basins. Peat represents
the first stage of the coalification series of peat →
lignite → bituminous coal → anthracite coal. The peat-
forming systems of the Okefenokee Swamp, Georgia, and Everglades
Swamp, Florida, were selected as suitable modern progenitors
of coal that are representative of past systems that eventually gave rise to huge accumulations of coal (Spackman *et al.*,
1974).

Contributions to the geology and botany of both peat-
forming areas are summarized elsewhere (Spackman *et al.*,
1974); however, a few comments on each system would be
appropriate. The Okefenokee Swamp is a paludal region of
swamps and marshes that lies on a Pleistocene terrace of the
Atlantic Coastal Plain in southeastern Georgia. This freshwater area of 1800 km^2 is the second largest swamp in the
United States. The Minnie's Lake area of the Okefenokee

(Casagrande *et al.*, 1977) is most representative of the swamp environment; it consists of a vegetational cover of *Taxodium*, *Magnolia* and *Nyssa* and has a peat depth of approximately 366 cm. As with many swamp environments (swamps cover 85% of the Okefenokee), the water cover in the Okefenokee is intermittent. The constantly water-covered environment of Grand Prairie is a good example of the marsh environment, which is the other major peat-forming system. The principal vegetation in this area is *Nymphaea*, *Utricularia* and *Orontium*; the depth of peat is approximately 400 cm. The pH in both areas is 4.0.

The Everglades environment is a region of 11,000 km^2 and is situated in southern Florida; the particular area of interest in the present study is in the saline peat-forming system along the Little Shark River; the principal vegetation is *Rhizophora Mangle L.* and the associated peat is 400-cm deep. This particular environment is only one of many sub-environments in the Everglades.

Peats from the Okefenokee and Everglades are viewed as the modern progenitors of low and high sulfur coals. Previous studies (Casagrande *et al.*, 1977; Casagrande and Siefert, 1977) showed that (1) the total amounts of sulfur found in coal (up to 10 percent on a dry basis) can be incorporated during the peat-forming stage, (2) all forms of sulfur found in coal can be found in peat, (3) environmental conditions such as salinity appear to have a dramatic effect on pyrite content, (4) total sulfur appears to increase with depth in the marine environment, (5) hydrogen sulfide is forming at all levels of peat, (6) the percentage of organic sulfur is similar to that reported in freshwater- and marine-derived coals, and (7) an ester-sulfate fraction accounts for 25 percent of the total sulfur associated with peat and appears to be an important contributor to the sulfur cycle.

In order to understand further the characterization of sulfur in coal progenitors, peat from the Grand Prairie (freshwater) and Little Shark (marine) were subjected to an organic fractionation. Thus peat samples from various levels of core, in each area, were fractionated to yield humic acid, fulvic acid, humin, water soluble and benzene/methanol soluble fractions. The distribution of organic and inorganic sulfur in each fraction was determined.

EXPERIMENTAL

Procedures used for coring and sample preservation are reported elsewhere (Casagrande and Erchull, 1976). Each peat sample selected from the various core levels in both environments (described above) was fractionated according to procedures described by Casagrande (1970). The fractionation resulted in the following fractions: water soluble, benzene/methanol soluble, humin, humic acid and fulvic acid.

The details of the procedure used for total sulfur analysis are reported elsewhere (Casagrande *et al.*, 1977); this involved an alkaline permanganate oxidation of all sulfur forms to sulfate and a subsequent reduction in a Johnson-Nishita apparatus (Johnson and Nishita, 1952) to hydrogen sulfide. After trapping the sulfur as zinc sulfide and developing a methylene blue color, the solution was measured spectrophotometrically; results were compared with standards (Tabatabai and Bremner, 1972).

The procedure used for inorganic sulfur determinations was a variation of that used for the analysis of total sulfur, with the oxidation step being eliminated. It was found (Freney, 1967) that the reducing solution consisting of hydriodic acid, hypophosphorous acid and formic acid reduces all inorganic sulfur forms in addition to ester-sulfate; carbon-bonded sulfur is not affected. Organic sulfur reported later represents the difference of total sulfur minus inorganic sulfur.

All analyses reported herein represent at least duplicate determinations; reproducibility was within 5 percent and many samples were cross-checked with independent analyses carried out by Harold Gluskoter at the Illinois State Geological Survey.

RESULTS AND DISCUSSION

Table I shows the distribution of organic matter as a function of peat fraction; data are presented for the freshwater-derived peat from Grand Prairie (Okefenokee) and marine-derived peat from the Little Shark River site (Everglades). In addition to this, Table I also displays data on the total peat sulfur associated with each peat fraction and the percentage of organic or inorganic sulfur associated with each fraction.

The major quantity of peat organic matter is in the humic fraction of both cores. This fraction contains hemicellulose, cellulose, lignin and kerogen. The first three polymers represent approximately 50 percent of the organic matter in the humic fraction; the remainder is in a kerogenous form. Humic acid, which represents the precursor to the vitrinitic material in coal, represents only a small proportion of the Grand Prairie peat, but humic acid in the Everglades represents almost 25 percent of the total organic matter. Since humic acid is a recognized diagenetic product, perhaps the conditions conducive to the process of humin → fulvic acid → humic acid are more prevalent in the mangrove (Everglades) environment than in the water lily marsh (Okefenokee) environment. This is understandable since the Little Shark area is subjected to tidal wash and consequently has a chance to be more oxidizing than the constantly water-covered marsh environment of Grand Prairie.

Table I

Distribution of Organic Matter and Sulfur in Peats

Sample	% of Total Organic Matter	% of Total Sulfur (d.w.b.)	% of Total Sulfur in Fraction	% of Total Sulfur as Organic	% of Total Sulfur as Inorganic
Okefenokee[a] *(Grand Prairie - fresh water)*					
Total	-	0.204	-	79.7	20.3
Water sol.	4.12	0.381	12.5	66.3	33.7
Ben/met sol.	4.78	0.151	1.56	60.2	39.8
Humin	69.2	0.164	67.8	78.3	21.7
Humic acid	5.83	0.329	13.2	70.5	29.5
Fulvic acid	16.1	0.048	4.47	35.2	64.8
Everglades[b] *(Little Shark - marine)*					
Total	-	1.66	-	59.4	40.6
Water sol.	2.05	5.92	7.29	0.83	99.1
Ben/met sol.	3.69	5.35	11.4	49.9	50.1
Humin	48.1	1.78	51.6	76.3	23.7
Humic acid	23.5	1.25	17.5	62.2	37.8
Fulvic acid	22.6	0.91	12.2	12.9	87.1

[a]Mean of five depths of peat in Grand Prairie freshwater peat core.
[b]Mean of five depths of peat in Little Shark marine water peat core.

While data are available from only one core in each
environment, some preliminary statements can be made about
the percentage of total sulfur in the fractions: (1) as
demonstrated in a previous study, the marine-derived peat has
more sulfur associated with it than the freshwater-derived
peat, (2) the content of sulfur in the water soluble fraction
probably reflects the contribution of sulfate to the peat,
(3) the sulfur associated with the benzene/methanol fraction
could be elemental sulfur, ester-sulfate and/or organic
sulfur, (4) humic acids appear to concentrate sulfur in
Grand Prairie environment but not in the mangrove environment,
(5) the humic content, which represents the major contributor
to the organic matter of the peat, has sulfur contents
similar to the total sample.

As a percentage of the total sulfur, the humic fraction
appears to contain most of the sulfur in peat by virtue of its
major contribution to the organic matter and not because it
concentrates sulfur. Humic acids appear to be second in the
amount of sulfur associated with fractions. The Okefenokee
humic acids have considerably higher levels of sulfur relative
to the total peat sulfur than the marine-derived Little Shark
peat. The precise reason for this is unclear at this time.
The benzene/methanol soluble fraction in the Everglades peat
makes a much larger contribution to the total sulfur than
the corresponding fraction in the Okefenokee peat; this may
be due to larger quantities of ester-sulfate in this fraction.

In comparing the percentage of organic vs inorganic
sulfur in the total peat, it is observed that most of the
sulfur is in an organic (*i.e.* C-S) combination. This is
probably due to contributions from sulfur-containing amino
acids. Also, some of the organic sulfur may result from
the interaction of hydrogen sulfide (which is being produced
at all levels) and aldehydes associated with the organic
matter. Presently experiments are being conducted to test
this idea. The fulvic acid fraction routinely has a higher
percentage of inorganic sulfur than organic sulfur associated
with it, probably because of the ester-sulfate content. The
sometimes large amounts of carbohydrates associated with this
fraction are excellent sites for ester-sulfate. Sulfate is
the main contributor to the relatively high inorganic sulfur
content of the water soluble fraction; this is especially
true for the marine Little Shark peat.

It appears that organic sulfur is widely distributed
within the peat fractions, very similar to observances in
coals. The organic sulfur content is certainly within the
realm of values obtained for both low and high sulfur coals.
It is difficult to believe that all the organic sulfur in
coal arose from amino acids, and thus studies presently being
undertaken (described above) will shed new light on this
problem. Preliminary results show a large uptake of H_2S

by the humin and humic acid fractions as a function of time.
Nissenbaum and Kaplan (1972) have demonstrated that sulfur
associated with marine humic acids is isotopically light
and thus may be of microbiological origin. This information
coupled with other ongoing studies on lignites will hopefully
result in a deeper understanding of sulfur incorporation in
the coal forming sequence.

ACKNOWLEDGMENTS

The authors thank Mr. Ganni Idowu and Mrs. Lily Ng
for their technical assistance in the laboratory. Drs.
William Spackman, Arthur Cohen and Peter Gunther are thanked
for their assistance in the field. Mr. John Eadie, Okefenokee
Refuge Manager is thanked for his close cooperation. This
work was supported by the Earth Science Section of the
National Science Foundation and Governors State University.

REFERENCES

Casagrande, D. J. and L. D. Erchull. "Metals in Peat from
Two Selected Areas in the Okefenokee Swamp," *Geochim.
Cosmochim. Acta* 40:387-393 (1976).

Casagrande, D. J. and K. Siefert. "Origins of Sulfur in
Coal: Importance of Ester-Sulfate Content of Peat,"
Science 195:675-676 (1977).

Casagrande, D. J., K. Siefert, C. Berschinski and N. Sutton.
Geochim. et Cosmochim. Acta 41:161-167 (1977).

Freney, J. R. "Sulfur-Containing Organics," in *Soil
Biochemistry* A. D. McLaren and G. H. Peterson, eds.,
(New York: Marcel Dekker, 1967), pp. 229-259.

Johnson, C. M. and H. Nishita. "Microestimation of
Sulfur," *Anal. Chem.* 24:736-742 (1952).

Niessenbaum, A. and J. R. Kaplan. "Chemical and Isotopic
Evidence for the *in situ* Origin of Marine Humic Substances,"
Limnol. Oceanog. 17(4):570-582 (1972).

Ruch, R. R., H. J. Gluskoter and N. F. Shimp. "Occurrence
and Distribution of Potentially Volatile Trace Elements
in Coal," *Ill. Geol. Surv. Environ. Geol. Notes.* 72
95 pp. (1974).

Spackman, W., A. D. Cohen, P. H. Given and D. J. Casagrande. *The Comparative Study of the Okefenokee Swamp and the Everglades-Mangrove Swamp-Marsh Complex of Southern Florida,* Field Guide Book for Geol. Soc. Amer. Meeting, Miami (1974).

Tabatabai, M. A. and J. M. Bremner. "Forms of Sulfur, Carbon, Nitrogen and Sulfur Relationships in Iowa Soils," *Soil Sci.* 114:380-386 (1972).

SECTION III

NITROGEN IN SOIL AND ITS IMPACT ON THE ATMOSPHERE

CHAPTER 41

NITROGEN TRANSFERS BETWEEN THE SOIL AND ATMOSPHERE

E. A. PAUL
R. L. VICTORIA*

Department of Soil Science
University of Saskatchewan
Saskatoon, Saskatchewan, Canada

INTRODUCTION

The major nitrogenous soil components are organic in nature. However, those most active with regard to plant nutrient uptake and transfers in and out of the soil are inorganic. On a long term basis in nonagricultural systems, transfers of N in and out of soil should be nearly equal. Over time, carbon shows a slight accumulation as shown by tundra, peat and fossil fuel deposits (Rosswall and Heal, 1975; Leith, 1975). Nitrogen should be expected to act similarly. However, man's cultivation and burning of virgin sites results in a drastic loss of soil organic N over extensive areas. Mineralization of residual C and N decreases logarithmically with time (Campbell *et al.*, 1976). Transfers, out of the soil on continued cultivation are not as large as initially but can still be substantial. Transfers to the atmosphere include NH_3 volatilization and the products of denitrification, NO, NO_2, N_2O and N_2. Other loss mechanisms include erosion and NO_3^- leaching.

NH_3 Volatilization

NH_3 volatilization plays a larger role than often thought. NH_3 escaping into the atmosphere has been estimated by Soderlund and Svensson (1976) to be: 2 Tg N/yr^{-1} from wild animals, 20-35 Tg from domestic animals, and 4-12 Tg from coal combustion, for a total of 26-53 Tg N/yr^{-1}. This estimate involves many assumptions. A technique for finding the NH_3 flux from animal wastes may be found in the $\delta^{15}N$ procedure.

*Permanent address: Centro de Energia Nuclear na Agricultura, Caixa Postal 96, 13400 Piracicaba, S. P., Brazil.

Generally plants have a $\delta^{15}N$ of +7 to 8 relative to atmospheric N, similar to that of soils (Wada *et al.*, 1975). Animals have slightly higher values, but manure exposed to the atmosphere has been found to range up to +24 (Kreitler, 1975). The heavier isotope (^{15}N) would be enriched by NH_3 volatilization and it should be possible to quantify the volatilization process by measuring this enrichment.

High NH_3 volatilization from soils coincides with high $CaCO_3$, alkaline pH values and low cation exchange capacity (Fenn, 1975; Mills *et al.*, 1974; van Veen, 1977). High temperatures and decreased moisture levels also result in higher volatilization (Clark *et al.*, 197 ; Fenn and Kessel, 1974; Lippold *et al.*, 1975; Ryan and Keeney, 1975; Stewart, 1970). Denmead *et al.*, (1976) noted that a grazed legume pasture lost up to 300 kg $N/ha^{-1}/day^{-1}$. An ungrazed area had a volatilization of 50 kg $N/ha^{-1}/day^{-1}$. Most of this large NH^3 flux, however, was reabsorbed by the vegetation and not lost from the system. However, NH_3 volatilization from leaf surfaces, especially dying leaves, also occurs (Martin and Ross, 1968; Porter *et al.*, 1972). Its significance has not been measured in detail.

Denitrification

Denitrification is the major process that releases N to the atmosphere. The microbiology and biochemistry of the processes involved have been reviewed (Broadbent and Clark, 1965; Delwiche and Bryan, 1976; Garcia, 1975a; Payne, 1973a; Stouthammer, 1976). Field and laboratory studies indicate that the major factors affecting this process in nature are O_2, a source of energy and the level of NO_3 (Bremner and Shaw, 1958; Garcia, 1975b). In aerobic soils, anaerobic microsites are required for biological denitrification. Nommik (1956) observed that denitrification decreased with increasing size of soil aggregates. Although O_2 tensions would be lower in the middle of large aggregates, NO_3^- and available C also would be decreased, leading to lower denitrification.

Soil water has its major influence through its effect on O_2 diffusion. It is generally agreed that denitrification is maximal at water saturation (Craswell and Martin, 1974; Stefanson, 1972a, 1972b). Below saturation, moisture continues to affect aeration. It also plays a general role in controlling plant growth and microbial activity.

Most denitrification is considered to be organotrophic in that carbonaceous substrates are involved (Payne, 1973a). However, inorganic substrates such as reduced sulfur compounds can play a major role under specific conditions (Mann *et al.*, 1972; Myers, 1972). The effect of organic matter is twofold. It is a substrate and the rate of organic matter decomposition affects the soil O_2 content. Several authors have established a linear relationship between soil organic C and denitrification (Bailey, 1976; Khan and Moore, 1968; Myers and McGarity, 1972; Smid and Beauchamp, 1976). Stanford *et al*., (1975) and Burford and Bremner (1975) reported good relationships between water soluble organic C and denitrification in soils. Because they excrete C from their roots and utilize CO_2, plants also tend to enhance denitrification (Bailey, 1976; Garcia, 1973, 1975b; Stefanson, 1976).

Temperatures and pH values affect microbial growth and thus denitrification. They also influence the products of denitrification in that processes favoring rapid denitrification produce mostly N_2, whereas acidic reactions and low temperatures result in varying but higher concentrations of N_2O (Cooper and Smith, 1963; Focht, 1974). Values for the NO_3 saturation in activated sludge systems (when denitrification is described by Michaelis-Menton kinetics) have been found to range from 0.05 to 0.06 ppm (Requa and Schroeder, 1973). However, Kohl *et al*., (1976) reported values of 4.6 to 48.7 ppm NO_3-N for nitrate saturation in soil systems. The difference between the sludge and the soil has been attributed to the difference in the diffusion rate of NO_3-N to the bacteria (van Veen, 1977).

Methods of Determining Denitrification Rates

Determination of the rate of denitrification from different systems has been limited by the availability of adequate methodology. Some of the systems studied and the methods utilized in studying these systems are summarized in Table I.

Table I
Approaches Used in Measuring Denitrification

Incubation Conditions	Methods of Measurement
Columns	^{15}N balance
Miscible displacement	$^{15}N_2$, $^{15}N_2O$, ^{15}NO accumulation
Lysimeters	N_2, N_2O, NO gas chromatography
Laboratory closed systems	NO_3^--NO_2^- depletion
Canopies	N_2O reduction
Field flux measurements	Infrared spectrometry
Respirometers	C_2H_2 blockage and N_2O accumulation
Gas lysimeters	
Sludge and water systems	

Nitrogen balance studies provide an indirect estimate of denitrification, especially when leaching losses can be accounted for (Allison, 1955; Delwiche and Bryan, 1976). These have led to a wide range of estimates, with an average of 15 percent of fertilizer N being unaccounted for. A recent C.A.S.T. (1976) publication used Hauck's (1969) estimate of $16/kg/ha^{-1}/yr^{-1}$ for cropland. Of this 1.5 kg was assumed to be N_2O (Focht, 1974). Noncultivated land was said to evolve 0.2 kg $N_2O/ha^{-1}/yr^{-1}$ for a total terrestrial N_2O flux of 4.2 Tg/yr^{-1}. Using a factor of $N_2:N_2O$ of 16:1, this would indicate a total denitrification of 71 Tg/yr^{-1}. This estimate should be compared to that of Soderlund and Svenssons (1976) of 69 Tg/hr^{-1} for N_2O alone to realize the discrepancies that still exist.

Under laboratory conditions, loss of NO_3^- or NO_2^- has often proven a useful measurement (Bowman and Focht, 1974; Dubey and Fox, 1974; Smid and Beauchamp, 1976; Stanford *et al.*, 1975). Measurements of the composition of the gaseous products evolved in closed systems provide a direct measurement of denitrification, infrared spectrophotometry for N_2O (Stevenson and Swaby, 1964). Mass spectrometry (Cady and Bartholomew, 1963; Nommik and Thorin, 1972) and gas chromatography (Burford and Stefanson, 1973; Dowdell and Smith, 1974; Ghoshal and Larsson, 1975; Payne, 1973b; Rolston *et al.*, 1976; Stefanson, 1972a, 1972b, 1972c) have been utilized with some degree of success. Difficulties arising in the analysis of the mixtures of N_2, N_2O, NO_2 and NO_2 enriched in ^{15}N make it hard to determine what fraction of each is contributing to the mass peak 30 during mass spectrometry.

The availability of good detectors and packing materials has made it possible to utilize gas chromatography, especially in systems where N_2 is absent in the initial gas phase (Balasubramanian and Kanehire, 1976; Burford and Bremner, 1975; Meed and Mackenzie, 1965). At normal atmospheric N_2 concentrations, present technology cannot detect small changes in N_2. Dowdell and Smith (1974) proposed that the measurement of N_2O in the soil profile under field conditions could be utilized as an index of denitrification. Actual denitrification occurring is difficult to calculate because of the uncertainty in the $N_2:N_2O$ ratio and their diffusion rates (Rolston *et al.*, 1976).

N_2O reduction has been utilized to measure denitrifying activity (Garcia, 1974, 1975c). The actual denitrifying activity was said to be obtained after 6-hr incubation, and the rate obtained after 20 hr represented the potential activity. The only criticism to this technique is that N_2O reduction is generally not the rate-limiting step in denitrification

In developing gas exchange assays for extra terrestrial life detection, Federorova *et al.*, (1973) noted that C_2H_2 blocked reduction of N_2O to N_2 by soil bacteria. Based on this finding, Balderston *et al.*, (1976) and Yoshinari and

Knowles (1976) tested C_2H_2 blockage of N_2O reduction as a method for determining denitrifying activity. Blockage was reversible when pure cultures of *P. perfectomarinus* were exposed to C_2H_2 concentrations up to 0.02 atm. At higher concentrations the blockage was not reversible; increasing concentrations of C_2H_2 increased the blockage of N_2O reduction. Yoshinari and Knowles (1976) observed that the presence of C_2H_2 did not affect NO_3^- and NO_2^- reduction by pure cultures, with negligible subsequent reduction of the N_2O produced. These data suggest the possibility of utilizing C_2H_2 blockage of N_2O reduction to measure denitrification under field conditions.

TESTING OF C_2H_2 BLOCKAGE

Before utilizing the above technique for field studies, it was necessary to test the extent of blockage and to quantify the reactions involved. This was done in our laboratory by verifying the blockage of N_2O reduction in two prairie soils. $^{15}NO_3$ additions then made it possible to study the reactions involved.

Methodology

Two soils were utilized: Annaheim, a black Chernozemic loam, had a pH of 6.8 and contained 26 μg/g NO_3-N, 2.5 μg/g NH_4^+ and 4 percent total C; Weyburn, a dark-brown Chernozemic, had a pH of 7.2 and contained 9.5 μg/g NO_3-N and 1.6 percent total C. The soil (100 g) was incubated at 20°C in 500-ml flasks with a side arm having a vacuum stopcock and a TS 14/35 joint. The flasks were stoppered with rubber stoppers containing a glass tube closed with a serum cap. Silicon sealant was around the stoppers and on the serum cap.

The flasks for anaerobic incubation were flushed by evacuation and back-filling with helium ten times. The final pressure in each flask was kept slightly above atmospheric. Flasks for aerobic incubation contained normal atmospheric air. Where required, C_2H_2 (10% v/v), 10 ml of a solution containing 2.26 mg NO_3-N, and H_2O sufficient to bring the moisture to 30% by weight were introduced with syringes through the serum caps.

The gas chromatograph was a 5710A Hewlett-Packard with a 15 uCi ^{63}Ni electron capture detector. A carrier gas of 95 percent argon and 5 percent methane had a 30-ml/min^{-1} flow rate through a 3.6 M x 3 mm stainless steel column packed with 50/80 mesh Poropak Q. The column temperature was 60°C, the detector temperature was 250° C, sample size was 1 ml. N_2O was eluted at 140 seconds after injection. The response was linear over a range of 0.2 ppm N_2O to 1100 ppm (volume basis).

At appropriate intervals gas samples were withdrawn with 1-ml syringes. At the termination of incubation, NH_4^+ and NO_3^- were extracted from the Annaheim soil with 2 N KCl at a 5:1 KCl soil solution. The Weyburn soil was extracted with water at a 1:5 soil-water ratio. NH_4^+ and NO_3^- were determined by steam distillation utilizing MgO and Devarda's alloy.

Total N was determined by Kjeldahl digestion to include NO_3^- and NO_2^- followed by steam distillation. ^{15}N was analyzed on a MAT GD150 mass spectrometer. All data were corrected for differences in volume among flasks.

Data Obtained

Figure 1 shows the N_2O-N evolution from 4.86 mg NO_3-N initially present in 100 g Annaheim soil during anaerobic incubation. N_2O evolution in the presence of C_2H_2 represents the mean of six flasks. The gas chromatographic analytical error amounted to 4-5 percent. The other portion of the standard deviations was attributed to differences in the denitrification rate among flasks. Denitrification proceeded rapidly and after four days the N_2O reached a plateau, indicating completion of the process. Incubation in the absence of C_2H_2 resulted in a small transitory N_2O peak, which was then reduced. Under aerobic conditions (Figure 2) in the presence of C_2H_2, N_2O was produced in the microgram range after a four-day lag as compared to N_2O production of milligrams of N_2O under anaerobic conditions without a lag period. A low but consistent level of N_2O production continued throughout the duration of the experiment. The N_2O evolution rate of 1 $\mu g(100$ g soil$)^{-1}$/day^{-1} would be equivalent to 4 kg/ha^{-1} over a six-month season.

The recovery of added plus native NO_3-N initially present in the soil after 16 days incubation of the Annaheim soil is shown in Table II. Seventy to 80 percent of the initial NO_3-N was recovered as N_2O. Small amounts of NH_4^+, NO_3^- and organic ^{15}N were also found, resulting in identification of 72-90 percent of the ^{15}N. The presence of a peak 30 and peak 31 during ^{15}N analysis of the gas phase indicated a small concentration of NO. The N_2 in the flask showed an abundance (A percent ^{15}N) of 0.5 percent in comparison to an initial $^{15}NO_3$ abundance of 2.767 percent (Table III). This indicated a significant concentration of N_2 in the atmosphere that did not come from the original nitrate. The percent NDNO$_3$ data indicate only 3.6-8 percent of the N_2 in the atmosphere was derived from the nitrate. The remainder is attributed to nonbiological processes such as leakage and possibly desorption of N_2 from the soil.

Analysis of the 28 and 29 peaks with the mass spectrometer and comparison with the peaks obtained by addition of standard amounts of N_2 to the same flasks at the termination

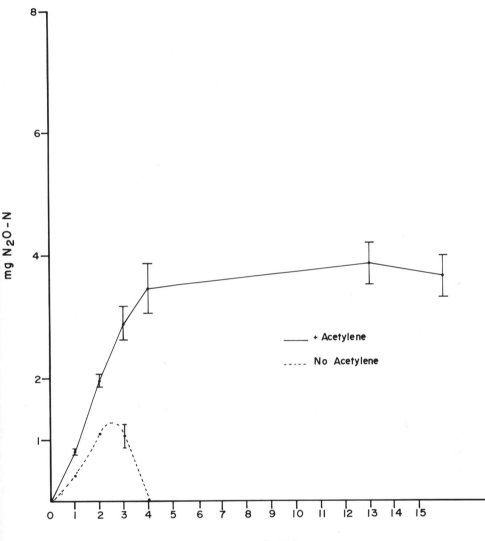

Figure 1. N_2O-N evolution from 4.86 mg NO_3^--N Annaheim soil
  anaerobic incubation.

of the experiment indicated a concentration of 0.2-0.5 percent
N_2.   If 5 percent of this was derived from $^{15}NO_3$, the $^{15}N_2$
in the flask accounted for less than 1 percent of the added
$^{15}NO_3$-N.

A higher percentage of initial NO_3-N was recovered in
the Weyburn soil after ten days anaerobic incubation in the
presence of C_2H_2 (Table IV).   The total recovery of initial
NO_3^--N ranged from 90-100 percent. $^{15}N_2$ present in the gas
phase after ten days anaerobic incubation (Table V) again
indicates a small evolution of $^{15}N_2$, accounting for a range

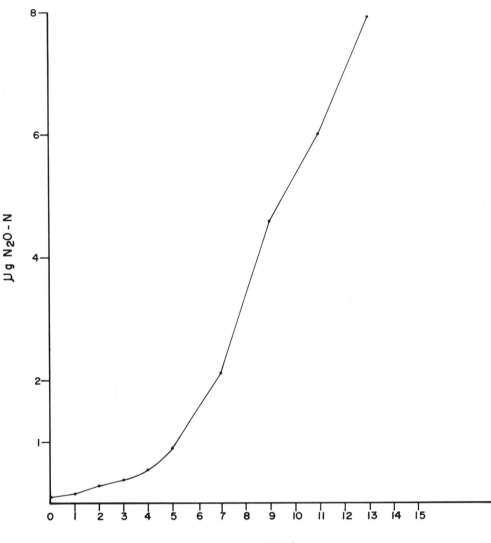

Figure 2.   N$_2$O evolution from 4.86 mg NO$_3^-$-N Annaheim soil
   aerobic incubation.

of 0.1-4.9 percent of the total N$_2$.   The 4.9 percent of the
N$_2$ derived from labelled nitrate (percent NDFNO$_3$) in flask 2
is inconsistent with the data in Table IV.   Quantitative
recovery was obtained from flask 2 without taking ^{15}N$_2$ into
consideration.

        A large difference in abundance between the N$_2$ and the
initial NO$_3^-$ again indicated that the source of the N$_2$ in the
flasks was not primarily via denitrification.   No mass 30
or 31 peak could be detected for the Weyburn soil.

Table II
Percent Recovery of NO_3-N
After Sixteen Days Incubation (Annaheim Soil)[a]

Flask	NH_4[b]	NO_3[c]	Soil[d]	N_2O-N[e]	Total
1	0.6	0.4	0.8	88	89.8
2	1.2	0.4	0.0	85	86.6
3	2.7	0.4	1.4	84	88.5
4	1.2	0.4	0.5	70	72.1
5	1.4	0.4	5.0	75	81.8
6	1.8	0.4	1.6	74	77.8

[a] 4.86 mg NO_3-N initially present in 100 g soil.
[b] Based on the difference between NH_4 after and before incubation.
[c] Based on the NO_3^- left after incubation.
[d] Based on the ^{15}N analysis of the soil after incubation and NH_4 + NO_3 extraction.
[e] Based on gas chromatographic data.

Table III
$^{15}N_2$ Present in the Gas-Phase
After Sixteen Days Incubation (Annaheim Soil)

Flask	A% $^{15}N_2$	%NDNO$_3$
1	0.4589	3.6
2	0.5066	5.6
3	0.5682	8.2
4	0.5430	7.1
5	0.5563	7.7
6	0.4851	4.7

Initial ^{15}N abundance of the nitrate = 2.767 A% ^{15}N.

Control ^{15}N abundance = 0.3729 A% ^{15}N.

% NDNO$_3$ = % N_2 derived from NO_3-N.

Table IV
Percent Recovery of NO_3-N
After Ten Days Incubation (Weyburn Soil)[a]

Flask	NO_3[b]	Soil[c]	N_2O[d]	Total
1	3.0	3.4	88	94.4
2	1.5	3.1	96	100.6
3	1.5	3.1	88	92.1
4	1.8	3.7	90	95.5

[a] 3.25 mg NO_3-N initially present in 100 g soil.
[b] Based on NO_3^- left after incubation.
[c] Based on ^{15}N analysis of the soil after incubation and NO_3^- extraction.
[d] Based on gas chromatographic data.

Table V
$^{15}N_2$ Present in the Gas-Phase
After Ten Days Incubation (Weyburn Soil).

Flask	A% $^{15}N_2$	% $NDFNO_3$
1	0.4227	0.2
2	1.4428	4.9
3	0.3853	0.1
4	0.3916	0.1

Initial NO_3^- ^{15}N abundance = 22.183 A% ^{15}N.

Control ^{15}N abundance = 0.3729 A% ^{15}N.

% $NDNO_3$ = % N_2 derived from NO_3-N.

DISCUSSION

The data indicate that the denitrification was largely blocked at the N_2O stage and further reduction was inhibited by the presence of C_2H_2. The ^{15}N data show a very small amount of $^{15}N_2$ evolved. The majority of the N_2 did not come from nitrate. The total N_2 concentrations in the flasks, however, were not high and it is considered that <1 percent of the original NO_3-N occurred as $^{15}N_2$. However, the data do indicate some lack of blockage in the heterogenous sample, such as soil. Work with pure cultures had indicated a complete blockage of N_2O production (Yoshinari and Knowles, 1976).

The evolution of $^{15}N_2$ (Tables III and V) also could be explained by chemodenitrification or reaction of NO_2^- with organic matter constituents (Stevenson *et al.*, 1970). This is indicated by the presence of NO in the Annaheim soil. It is possible that some bacteria that do not include N_2O as an intermediate in the denitrification pathway as suggested by Cook *et al.*, (1973) also exist within the soil sample.

Quantification of the $^{15}N_2$ and NO evolved requires further detailed mass spectrometric measurements to overcome problems with leakage and degradation of N_2O in the mass spectrometer. The data we have obtained, however, do suggest that blockage of denitrification by C_2H_2 and measurement of the N_2O evolved with a sensitive gas chromatograph could be an excellent technique for determination of denitrification.

The two soils utilized showed differing rates of N_2O production under anaerobic conditions. Both soils contained adequate available organic C to cause complete denitrification of the initial NO_3-N. The higher organic matter Annaheim soil had higher denitrification rates. It showed a capacity for evolving 100 μg N/g of soil during the active denitrification phase before NO_3 became limiting.

The sensitivity of the technique is further exemplified by the aerobic incubations of the Annaheim soil where microgram concentrations of N_2O were produced by 100 g soil after an initial lag phase. These incubation conditions approximate those that would be obtained under field conditions with a normal atmosphere and moisture content. It is envisioned that this technique will make it possible to measure denitrification rates under normal field conditions during various times of the year.

ACKNOWLEDGMENTS

The research in this paper was conducted by R. L. Victoria while at the University of Saskatchewan through funds provided by the International Atomic Energy Agency, Vienna.

REFERENCES

Allison, F. E. "The Enigma of Soil Nitrogen Balance Sheets," *Adv. Agron.* 7:213-250 (1955).

Bailey, L. D. "Effects of Temperature and Root on Denitrification in a Soil," *Can. J. Soil Sci.* 56:79-87 (1976).

Balasubramanian, V. and Y. Kanehiro. "Denitrification Potential and Pattern of Gaseous N Loss in Tropical Hawaiian Sils," *Trop. Agric. (Trinidad)* 53:293-303 (1976).

Balderston, W. L., B. Sherr and W. J. Payne. "Blockage by Acetylene of Nitrous Oxide Reduction in *Pseudomonas perfectomarinus*," *Appl. Environ. Microbiol.* 31:504-508 (1976).

Bowman, R. A. and D. D. Focht. "The Influence of Glucose and Nitrate Concentrations Upon Denitrification Rates in Sandy Soils," *Soil Biol. Biochem.* 6:297-301 (1974).

Bremner, J. M. and K. Shaw. "Denitrification in Soil. II. Factors Affecting Denitrification," *J. Agric. Sci.* 51:40-52 (1958).

Broadbent, F. E. and F. E. Clark. "Denitrification," in *Soil Nitrogen* W. F. Bartholomew and F. E. Clark, eds., (Madison: American Society of Agronomy, 1965).

Burford, J. R. and J. M. Bremner. "Relationships Between the Denitrification Capacities of Soils and Total, Soluble and Readily Decomposable Soil Organic Matter," *Soil Biol. Biochem.* 7:389-394 (1975).

Burford, J. R. and R. C. Stefanson. "Measurement of Gaseous Losses of Nitrogen from Soils," *Soil Biol. Biochem.* 5:133-141 (1973).

Cady, F. B. and W. V. Bartholomew. "Investigations of Nitric Oxide Reactions in Soils," *Soil Sci. Soc. Amer. Proc.* 27:546-549 (1963).

Campbell, C. A., E. A. Paul and W. B. McGill. "Nitrogen Contents of Soils of the Canadian Prairies," in *Proceedings of the Western Canada Nitrogen Symposium*, Calvary, (1976).

Clark, F. E., C. V. Cole and R. A. Bowman. "Nutrient Cycling," in *Grassland Systems Analysis and Mankind*, A. I. Breymeyer and G. M. Van Dyne, eds. (Cambridge: Cambridge University Press).

Cook, F. D., R. P. Wellman and H. R. Krouse. "Nitrogen Isotope Fractionation in the Nitrogen Cycle," in *Proceedings of the Symposium on Hydrogeochemistry and Biogeochemistry*, Vol. II:49-64 (1973).

Cooper, G. S. and R. L. Smith. "Sequence of Products Formed During Denitrification in Some Diverse Western Soils," *Soil. Sci. Soc. Amer. Proc.* 27:659-662 (1963).

Council for Agricultural Science and Technology. *Effect of Increased Nitrogen Fixation on Stratospheric Ozone*, 53 (1976).

Craswell, E. T. and A. E. Martin. "Effect of Moisture Content on Denitrification in a Clay Soil," *Soil Biol. Biochem.* 6:127-129 (1974).

Delwiche, C. C. and B. A. Bryan. "Denitrification," *Ann. Rev. Microbiol.* 30:241-262 (1976).

Denmead, O. T., J. R. Freney and J. R. Simpson. "A Closed Ammonia Cycle Within a Plant Canopy," *Soil. Biol. Biochem.* 8:161-164 (1976).

Dowdell, R. J. and K. A. Smith. "Field Studies of the Soil Atmosphere. II. Occurrence of Nitrous Oxide," *J. Soil Sci.* 25:231-238 (1974).

Dubey, H. D. and R. H. Fox. "Denitrification Losses From Humid Tropical Soils of Puerto Rico," *Soil Sci. Soc. Amer. Proc.* 38:917-120 (1974).

Federova, R. I., E. I. Milekhina and N. I. Ilyukhina. "Possibility of Using the 'Gas-Exchange' Method to Detect Extra Terrestrial Life: Identification of Nitrogen Fixing Organisms," *Akad. Nauk. SSR Izrestia Ser. Biol.* 6:797-806 (1973).

Fenn, L. B. "Ammonia Volatilization From Surface Applications of Ammonium Compounds on Calcareous Soils, III. Effects of Mixing Low and High Loss Ammonium Compounds," *Soil Sci. Soc. Amer. Proc.* 39:366-368 (1975).

Fenn, L. B. and D. E. Kissel. "Ammonia Volatilization From Surface Applications of Ammonium Compounds on Calcareous Soils. II. Effects of Temperature and Rate of NH_4^+-N Application," *Soil. Sci. Soc. Amer. Proc.* 38:606-610 (1974).

Focht, D. D. "The Effect of Temperature, pH and Aeration on the Production of Nitrous Oxide and Gaseous Nitrogen- A Zero-Order Kinetic Model," *Soil Sci.* 118:173-179 (1974).

Garcia, J. L. "Influence de la Rhizosphere du riz sur l'activité dénitrifiante potentielle des sols de rizières du Sénégal," *Ecol. Plant.* 8:315-323 (1973).

Garcia, J. L. "Reduction de l'oxyde nitreux dans les sols de rizières du Sénégal: Mesure de l'activité dénitrifiante," *Soil Biol. Biochem.* 6:79-84 (1974).

Garcia, J. L. "La dénitrification dans les sols," *Bulletin de l'institut Pasteur* 73:167-193 (1975a).

Garcia, J. L. "Effet rhizosphère du riz sur la dénitrification," *Soil Biol. Biochem.* 7:139-141 (1975b).

Garcia, J. L. "Evaluation de la dénitrification dans les rizières par la méthode de reduction de N_2O. *Soil Biol. Biochem.* 7:251-256 (1975c).

Ghoshal, S. and B. Larsson. "Gas Chromatographic Studies on Soil Denitrification. I. An Isothermic Separation Method for Evolved Gases Using Carbosieve-B as the Gas Chromatographic (GC) Packing," *Acta Agriculturae Scandinavica* 25:275-270 (1975).

Hauck, R. D. "Quantitative Estimates of N Cycle Processes: Review and Comments," in *Recent Developments in the Use of ^{15}N in Soil-Plant Studies*, sponsored by the Joint FAO/IAEA, Division of Atomic Energy in Food and Agriculture Sofia, Bulgaria (1969).

Khan, M. F. A. and A. W. Moore. "Denitrifying Capacity of Some Alberta Soils," *Can. J. Scol Sci.* 48:89-91 (1968).

Kohl, D. H., F. Vithayathil, P. Whitlow, G. Shearer and S. H. Chien. "Denitrification Kinetics in Soil Systems: The Significance of Good Fits of Data to Mathematical Forms," *Soil Sci. Soc. Amer. J.* 40:249-253 (1976).

Kreitler, C. W. *Determining the Source of Nitrate in Ground Water by Nitrogen Isotope Studies*, Bureau of Economic Geology, Austin, Report of Investigations 83 (1975).

Leith, H. "Primary Production of the Major Vegetation Units of the World," in *Primary Productivity of the Biosphere* H. Leith and R. H. Whittaker eds., (Berlin: Springer-Verlag, 1975).

Lippold, H., R. Heber and I. Forster. "Ammoniakverluste bei Harnstoffdüngung. I. Mitteilung: Modellversuche zur Ammoniakverflüchtigung in Abhängigkeit von pH-Wert, Austauschkapazität, Temperatur und Wassergehalt des Boden," *Arch. Acker- u. Pflanzenbau u. Bodenkd.* 19:619-630 (1975).

Mann, L. D., D. D. Focht, H. A. Joseph and L. H. Stolzy. "Increased Denitrification in Soils by Additions of Sulfur as an Energy Source," *J. Environ. Quality* 1:329-332 (1972).

Martin, A. E. and P. J. Ross. "A Nitrogen Balance Study Using Labelled Fertilizer in a Gas Lysimeter," *Plant and Soil* 28:182-186 (1968).

Meek, B. D. and A. J. MacKenzie. "The Effect of Nitrite and Organic Matter on Aerobic Gaseous Losses of Nitrogen from a Calcareous Soil," *Soil Sci. Soc. Amer. Proc.* 29:176-178 (1965).

Mills, H. A., A. V. Barker and D. N. Maynard. "Ammonia Volatilization from Soils," *Agron. J.* 66:355-358 (1974).

Myers, R. J. K. "The Effect of Sulphide on Nitrate Reduction in Soil," *Plant and Soil* 37:431-433 (1972).

Myers, R. J. K. and J. W. McGarity. "Denitrification in Undisturbed Cores from a Solodized Solonetz B Horizon," *Plant and Soil* 37:81-89 (1972).

Nommik, H. "Investigations on Denitrification in Soils," *Acta Agric. Scand.* 6:195-228 (1956).

Nommik, H. and J. Thorin. "Transformations of ^{15}N-Labelled Nitrite and Nitrate in Forest Raw Humus During Anaerobic Incubation," in *Isotopes and Radiation in Soil-Plant Relationships Including Forestry* I.A.E.A. Vienna (1972).

Payne, W. J. "Reduction of Nitrogenous Oxides by Micro-organisms," *Bacteriol. Rev.* 37:409-452 (1975a).

Payne, W. J. "The Use of Gas Chromatography for Studies of Denitrification in Ecosystems," *Bull. Ecol. Res. Comm. Stockholm* 17:263-268 (1975b).

Porter, L. K., F. Viets and G. L. Hutchinson. "Air Containing Nitrogen-15 Ammonia: Foliar Absorption by Corn Seedlings," *Science* 175:759-761 (1972).

Requa, D. A. and E. D. Schroeder. "Kinetics of Packed-Bed Denitrification," *J. Water Poll. Contr. Fed.* 45:1696-1707 (1973).

Rolston, D. E., M. Fried and D. A. Goldhamer. "Denitrification Measured Directly from Nitrogen and Nitrous Oxide Gas Fluxes," *Soil Sci. Soc. Amer. J.* 40:259-266 (1976).

Rosswall, T. and O. W. Heal. *Structure and Function of Tundra Ecosystems* Ecological Bulletin, Swedish Natural Science Research Council, Stockholm (1975).

Ryan, J. A. and D. R. Keeney. "Ammonia Volatilization from Surface-Applied Waste Water Sludge," *J. Water Poll. Contr. Fed.* 47:386-393 (1975).

Smid, A. E. and E. G. Beauchamp. "Effects of Temperature and Organic Matter on Denitrification in Soil," *Can. J. Soil Sci.* 56:385-391 (1976).

Soderlund, R. and B. H. Svensson. "The Global Nitrogen Cycle," in *Nitrogen, Phosphorus and Sulphur-Global Cycles* B. H. Svensson and R. Soderlund eds., SCOPE Report 7 Ecol. Bull. (Stockholm), pp. 23-73 (1976).

Stanford, G., R. A. Vander Pol and S. Dzienia. "Denitrification Rates in Relation to Total and Extractable Soil Carbon," *Soil Sci. Soc. Amer. Proc.* 39:284-289 (1975).

Stefanson, R. C. "Soil Denitrification in Sealed Soil-Plant Systems. I. Effect of Plants, Soil Water Content and Soil Organic Matter Content," *Plant and Soil* 37:113-127 (1972a).

Stefanson, R. C. "Soil Denitrification in Sealed Soil-Plant Systems. II. Effect of Soil Water Content and Form of Applied Nitrogen," *Plant and Soil* 37:129-140 (1972b).

Stefanson, R. C. "Soil Denitrification in Sealed Soil-Plant Systems. III. Effect of Disturbed and Undisturbed Soil Samples," *Plant and Soil* 37:141-149 (1972c).

Stefanson, R. C. "Denitrification from Nitrogen Fertilizer Placed at Various Depths in the Soil-Plant System," *Soil Sci.* 121:353-363 (1976).

Stevenson, F. J., R. M. Harrison, R. Wetselaar and R. A. R. A. Leeper. "Nitrosation of Soil Organic Matter: III. Nature of Gases Produced by Reaction of Nitrite with Lignins, Humic Substances, and Phenolic Constituents under Neutral and Slightly Acidic Conditions," *Soil Sci. Soc. Amer. Proc.* 34:430-435 (1970).

Stevenson, F. J. and R. J. Swaby. "Nitrosation of Soil Organic Matter: I. Nature of Gases Evolved During Nitrous Acid Treatment of Lignins and Humic Substances," *Soil Sci. Soc. Amer. Proc.* 28:773-778 (1964).

Stouthamer, A. H. "Biochemistry and Genetics of Nitrate Reductase in Bacteria," *Adv. Microbiol.* 14:315-375 (1976).

Stewart, B. A. "Volatilization and Nitrification of Nitrogen Under Simulated Feed Lot Conditions," *J. Environ. Sci. Tech.* 4:579-582 (1970).

van Veen, J. A. "The Behaviour of Nitrogen in Soil, A Computer Simulation Model," *Ph.D. Thesis* ITAL, Wageningen (1977).

Wada, E., T. Kodonaga and S. Matsuo. "^{15}N Abundance in Nitrogen of Naturally Occurring Substances and a Global Assessment of Denitrification from Isotopic Viewpoints," *Geochem. Journ.* 9:139-148 (1975).

Yoshinari, T. and R. Knowles. "Acetylene Inhibition of Nitrous Oxide Reduction by Denitrifying Bacteria," *Biochem. Biophys. Res. Comm.* 69:705-710 (1976).

CHAPTER 42

NITROGEN FIXATION AND DENITRIFICATION IN ARID SOIL CRYPTOGAMIC CRUST MICROENVIRONMENTS

J. SKUJINS
B. KLUBEK

Ecology Center
Utah State University
Logan, Utah 84322 USA

INTRODUCTION

Nitrogen fixation by desert algal crusts and denitrification probably constitutes the major input and loss of nitrogen into and out of desert ecosystems (Mayland, McIntosh and Fuller, 1966; Rychert, 1975; Rychert and Skujins, 1974; Tucker and Westerman, 1973; Westerman and Tucker, 1977). The nitrogen fixed by the algal crusts is available to desert plants (Mayland and McIntosh, 1966); however, evidence suggests that most of the nitrogen fixed by algal crust is lost by denitrification (Skujins and West, 1974). Only a few studies have been made to assess the environmental and ecological parameters affecting biological nitrogen fixation and denitrification (Rychert *et al.*, 1977; Skujins and West, 1974) and no attempt has been made to investigate both of these biological processes simultaneously, but rather *sui generis*. Most of the studies on biological nitrogen fixation have been with more temperate ecosystems (Hardy, Burns and Holsten, 1973; Hardy *et al.*, 1968), and denitrification has been associated with agricultural systems (Alexander, 1967; Allison, 1955, 1966).

Part of the research launched by the U.S.I.B.P. Desert Biome was aimed at unraveling the nitrogen cycle in several examples of American arid areas. In this report we present some results regarding nitrogen processes involving N_2-fixing cyanophyte phycosphere in soil crusts.

METHODOLOGY

Conventional methods were used for chemical and biological analysis of soils (Skujins, 1972). The 15-N technique was used for tracing the nitrogen cycle, including denitrification (Skujins and West, 1974). The research site was located

in the Great Basin Desert, Curlew Valley, Utah. The average
annual precipitation in the area is 19.5 cm, about half of
which appears as snow. The soils are clay-containing, silt-
loam, calciorthids and have been described elsewhere
(Skujins, 1973).
     Much of the clay-containing soil surface is covered with
a crust consisting mostly of lichens and blue-green algae.
About 20 percent of the blue-green algae found in the crust
and in the association with lichens have nitrogen-fixing
ability.

RESULTS AND DISCUSSION

Nitrogen Fixation

     The average N_2-fixation by surface blue-green algal-
lichen cover from spring rains was 0.008 kg N/ha/hr, and the
variation ranged from 0.005 to 0.010 kg N/ha/hr. The variation results from the nonuniformity of the crust in the field.
     An estimate of annual nitrogen fixation by the algal/
lichen crust may be made, based primarily upon the fall and
spring rainy periods *in situ*. Approximately 120 days of
nitrogen fixation, a minimum of 10 kg N fixed/ha/yr to a
maximum of about 100 kg N fixed/ha/yr, may be estimated for
areas of heavy lichen cover. Absolute amounts would vary
with weather, plant community and extent of algal-lichen cover.
An average value for Curlew Valley site was estimated 16 kg
N/ha/yr.

Denitrification

     The major loss of nitrogen from this type of an arid
ecosystem is by denitrification. Our data indicate that
nearly all of the N_2 fixed by the algal-lichen crust is
rapidly denitrified over a period of three weeks.
     Table I shows the $^{15}N_2$ fixation data by the Curlew
Valley soil crusts. The *Ceratoides* site crust had fixed the
greatest amount of ^{15}N, while the *Atriplex* crust had fixed
the least amount of ^{15}N per given conditions.

Table I
$^{15}N_2$ Fixation by Soil Crust, Curlew Valley, Utah

Dominant Vegetation	μg $^{15}N_2$ Fixed/g Soil/ 10 cm^2/24 hr Mean ±S.D.
Artemisia	1.33 (±0.18)
Ceratoides	5.15 (±0.11)
Atriplex	0.41 (±0.04)

Table II gives the ^{15}N analysis of the crusts following a postfixation incubation in the dark. It is observed that most of the fixed ^{15}N resides in the organic fraction, while detectable levels of ^{15}N in the NO_2^- and NO_3^- fraction were

Table II
^{15}N Flux in $^{15}N_2$-Exposed Soil Crusts After Seven Days Incubation

^{15}N Fraction	µg ^{15}N/g Soil		% Loss ^{15}N
	Mean	±S.D.	
Artemisia site crust			
Organic ^{15}N	0.69	(±0.003)	
$^{15}NH_4^+$	0.13	(±0.01)	
$^{15}NO_2^- + {}^{15}NO_3^-$	0.0		
Total ^{15}N recovered	0.82	(±0.07)	38.4
Initial total ^{15}N	1.33	(±0.18)	
Ceratoides site crust			
Organic ^{15}N	4.07	(±0.31)	
$^{15}NH_4^+$	0.29	(±0.03)	
$^{15}NO_2^- + {}^{15}NO_3^-$	0.0		
Total ^{15}N recovered	4.36	(±0.28)	15.3
Initial total ^{15}N	5.15	(±0.11)	
Atriplex site crust			
Organic ^{15}N	0.23	(±0.001)	
$^{15}NH_4^+$	0.10	(±0.011)	
$^{15}NO_2^- + {}^{15}NO_3^-$	0.0		
Total ^{15}N recovered	0.33	(±0.013)	19.5
Initial total ^{15}N	0.41	(±0.04)	

nil. However, the *Artemisia* soil curst experienced the greatest ^{15}N loss due to denitrification (38.4 percent), while the *Ceratoides* soil crust had the least (15.3 percent). Table III summarizes the recovery of ^{15}N following a 10-week post^{15}N-fixation exposure. The data show that less than 1 percent of the fixed $^{15}N_2$ was recovered, indicating denitrification losses.

The data indicate that the denitrification process is quite rapid. To verify the data, other sources of ^{15}N were applied and the present recovery was determined after a three-week incubation period.

Table III
Recovery of ^{15}N in the Various N Fractions
Ten Weeks After ^{15}N Fixation by Algal/Lichen Crust

Fraction	Atom % Excess ^{15}N	µg ^{15}N
Crust at initial time	1.064	933
Crust after 10 weeks:		
Exchangeable NH_4^+	0.039	0.079
$NO_2^- + NO_3^-$	0.014	0.129
Fixed NH_4^+	0.007	0.260
Organic N	0.003	1.810

Total recovery after 10 weeks <1 percent.

Table IV lists the soil ^{15}N analysis of $(^{15}NH_4)_2SO_4$ treated *Artemisia* soil. There was an increase in $^{15}NO_3^-$ followed by a decline. Exchangeable $^{15}NH_4^+$ was initially low; an increase and then a decline were noted. With slight variations, similar processes took place in *Atriplex* and *Ceratoides* site soils. In the three soils, $^{15}NO_2^-$ did not appear in detectable amounts. The weekly ^{15}N denitrification loss for the *Artemisia* soil was 60.9, 67.4 and 73.5 percent; the *Ceratoides* soil, 50.7, 61.4 and 77.4 percent; the *Atriplex* soil, 64.3, 66.9 and 84.4 percent. In all cases, the ^{15}N loss due to denitrification increased with time. The addition of dried plant material to the $(^{15}NH_4)_2SO_4$ amended soils also influenced the loss of applied ^{15}N. In soils amended with plant material (Table V), the highest losses were from 75.7 to 85.4 percent. The plant litter-amended soils had somewhat lower denitrification losses of applied $(^{15}NH_4)_2SO_4$, ranging from 64 to 80 percent.

The loss of nitrogen was also assessed utilizing $K^{15}NO_3$ (Table VI); 56.2 to 67.5 percent of the applied $^{15}NO_3^-$ was denitrified. Some of the remaining ^{15}N (5 to 9.1 µg) resides in either NH_4^+ or organic nitrogen form.

In most of the experimental systems, 60 percent or more of the biologically fixed $^{15}N_2$, and of applied $(^{15}NH_4)_2SO_4$ and $K^{15}NO_3$ was lost due to denitrification. Such processes may readily occur in the algal crust when rain, snowmelt, or dew activates the crust, thus allowing rapid fixation of both carbon and nitrogen. In the Curlew Valley soil crusts, two algal dominants, *Lyngbya* and *Microcoleus*, have been identified (Sorensen, 1975), and Brock (1975) has described bacterial populations associated with such crusts as well. *Lyngbya* can fix both nitrogen and carbon, while *Microcoleus* can only fix carbon. Thus a blue-green algae-bacteria association in the phycosphere appears as an important possibility. As nitrogen is fixed and eventually released into soil in the

Table IV
Surface Soil ^{15}N Fractions from $(^{15}NH_4)_2SO_4$ Amended,
Artemisia Site Air-Drying Soil after 21 Days.
Initial ^{15}N Content = 912 μg

N Fraction	μg ^{15}N/100 g soil Mean	±S.D.	Atom % excess ^{15}N	% Loss ^{15}N
Time: 7 Days				
Exchangeable $^{15}NH_4^+$-N	8.8	(±0.97)	0.966	
Fixed $^{15}NH_4$-N	15.9	(±0.16)	0.265	
Organic ^{15}N	70.84	(±2.2)	0.088	
$^{15}NO_2^-$-N	0.0		0.0	
$^{15}NO_3^-$-N	249.1	(±3.09)	15.47	
$^{15}NH_3$-N (volatilized)	11.55		---	
Total	356.2	(±2.2)	---	60.9
Time: 14 Days				
Exchangeable $^{15}NH_4^+$-N	20.03	(±5.9)	3.18	
Fixed $^{15}NH_4^+$-N	11.30	(±0.48)	0.178	
Organic ^{15}N	59.52	(±5.85)	0.077	
$^{15}NO_2^-$-N	0.0		0.0	
$^{15}NO_3^-$-N	206.32	(±8.11)	10.69	
$^{15}NH_3$-N (volatilized)	12.61		---	
Total	297.3	(±5.4)	---	67.4
Time: 21 Days				
Exchangeable $^{15}NH_4^+$-N	5.18	(±4.11)	0.673	
Fixed $^{15}NH_4^+$-N	20.48	(±1.18)	0.318	
Organic ^{15}N	77.93	(±7.33)	0.094	
$^{15}NO_2^-$-N	0.00		0.00	
$^{15}NO_3^-$-N	138.43	(±8.65)	6.18	
$^{15}NH_3$-N (volatilized)	12.8		---	
Total	242.0	(±17.7)	---	73.5

Table V
Surface Soil ^{15}N Fraction
from (^{15}NH$_4$)$_2$SO$_4$ + Plant Material Amended,
Air-Drying Soils After 21 Days.
Initial ^{15}N Content = 912 µg

N fraction	µg N/100 g soil Mean	±S.D.	Atom % Excess ^{15}N	% Loss ^{15}N
Artemisia site soil	8.17	(±0.88)	1.22	
Exchangeable ^{15}NH$_4$$^+$-N	39.47	(±0.49)	0.51	
Fixed ^{15}NH$_4$$^+$-N	79.97	(±0.74)	0.074	
Organic ^{15}N	0.62	(±0.41)	2.07	
^{15}NO$_2$$^-$-N	5.95	(±0.7)	3.72	
^{15}NO$_3$$^-$-N	7.95	(±6.34)	--	
^{15}NH$_3$-N (volatilized)	132.8	(±15.3)	--	85.4
Total				
Ceratoides site soil				
Exchangeable ^{15}NH$_4$$^+$-N	6.96	(±1.18	1.07	
Fixed ^{15}NH$_4$$^+$-N	13.22	(±0.82)	0.157	
Organic ^{15}N	91.67	(±1.99)	0.089	
^{15}NO$_2$$^-$-N	1.39	(±0.55)	0.24	
^{15}NO$_3$$^-$-N	83.0	(±10.1)	1.52	
^{15}NH$_3$-N (volatilized)	25.09	(±3.23)	--	
Total	221.3	(±3.23)	--	75.7
Atriplex site soil				
Exchangeable ^{15}NH$_4$$^+$-N	2.82	(±1.06)	.706	
Fixed ^{15}NH$_4$$^+$-N	21.0	(±1.31)	0.298	
Organic ^{15}N	52.7	(±5.63)	0.064	
^{15}NO$_2$$^-$-N	0.0		0.0	
^{15}NO$_3$$^-$-N	45.5	(±7.95)	1.06	
^{15}NH$_3$-N (volatilized)	64.81	(±23.3)	--	
Total	186.8	(±6.07)	--	79.5

Table VI
Reduction of Added $K^{15}NO_3$
Following a Seven Day Incubation Period
of Curlew Valley Surface Soils

| N Fraction | μg ^{15}N/g soil | | % Loss ^{15}N |
	Mean	±S.D.	
Artemisia site	9.10	(±0.52)	
Organic + $NH_4{}^+$-^{15}N	3.34	(±0.23)	
$^{15}NO_2{}^-$	29.97	(±2.39)	
$^{15}NO_3{}^-$	42.41	(±2.22)	
Total recovered	96.8		56.2
Initial $^{15}NO_3{}^-$			
Ceratoides site			
Organic + $NH_4{}^+$-^{15}N	5.22	(±0.52)	
$^{15}NO_2{}^-$	5.08	(±0.34)	
$^{15}NO_3{}^-$	21.20	(±10.0)	
Total recovered	31.5	(±9.8)	
Initial $^{15}NO_3{}^-$	96.8		67.5
Atriplex site			
Organic + $NH_4{}^+$-^{15}N	4.98	(±0.01)	
$^{15}NO_2{}^-$	6.19	(±0.75)	
$^{15}NO_3{}^-$	26.03	(±2.4)	
Total recovered	37.2	(±3.08)	
Initial $^{15}NO_3{}^-$	96.8		61.6

form of $NH_4{}^+$, it becomes available for nitrification within a few hours (Skujins and West, 1974).

The alternate wet and dry cycles of the desert enhance the nitrification process, and the carbon, fixed by the crust, is a ready energy source for denitrifying organisms. As the algal crust is moistened, the soil respiration rate rapidly increases and the oxygen tension drops significantly. Denitrification may occur at oxygen concentration of about 1 percent and below. At these intermediate conditions, the thermodynamically most probable form of gaseous N loss is N_2O. Other sutdies by Skujins and West (1974) on the nitrogen processes in the cryptogamic crust have shown N_2O to be the predominant form of N released into the atmosphere. This loss is rapid, and nearly 20 percent may be lost in a week, or as much as 80 percent in five weeks and 99 percent in an annual season.   Thus, although the cryptogamic crust may fix large quantities of N_2, little nitrogen is added to the ecosystem due to the denitrification process.

CONCLUSIONS

During the snowmelt and rainy seasons of spring and autumn, the blue-green algae, a component of the soil crypto-gamic crust, exhibit high activity, not only in N_2 fixation but also as primary producers.   The maximum activity values at comparatively low temperature, low light intensity and high moisture content are significant, since no algal activity may take place on the soil surface upon the drying of soils at elevated temperatures in direct sunlight.   During the summer, nitrogen fixation is associated with lower tempera-tures in the morning and the presence of dew (Rychert and Skujins, 1974).

The fixed nitrogen is released in the soil as ammonium, where most of it is nitrified to nitrite (NO_2^-) or nitrate (NO_3^-), then again reduced to N_2 or N_2O, and lost to the ecosystem.   The easily degradable organic matter, resulting from the primary production by algae, is available as the energy source for denitrification.   The process implies that the oxidation of ammonia to nitrite and the reduction of nitrite to nitrogen gas takes place in the same microenviron-ment at the same time, most of it in the phycosphere.   The whole cycle may thus be completed within hours under favorable environmental conditions.

ACKNOWLEDGMENTS

We wish to thank Dr. Paul Eberhardt for significant contributions to this project.   This investigation was supported in part by the US IBP Desert Biome, National Science Foundation Grant No. GB-15886.

REFERENCES

Alexander, M.   *Introduction to Soil Microbiology* (New York: John Wiley and Sons, 1967).

Allison, F. E.   "The Enigma of Soil Nitrogen Balance Sheets," *Adv. Agron.* 7:213-250 (1955).

Allison, F. E.   "The Fate of Nitrogen Applied to Soils," *Adv. Agron.* **18**:219-258 (1966).

Brock, T. D. "Effect of Water Potential on a *Microcoleus* Cyanophyceae) from a Desert Crust," *J. Phycol.* 11:316-320 (1975).

Hardy, R. W. F., R. D. Holsten, E. K. Jackson and R. C. Burns. "The Acetylent-Ethylene Assay for N₂ Fixation: Laboratory and Field Evaluation," *Pl. Physiol.* 43:1185-1207 (1968).

Hardy, R. W. F., R. C. Burns and R. D. Holsten. "Application of the Acetylene Reduction Assay for Measurement of Nitrogen Fixation," *Soil Biol. Biochem.* 5:47-81 (1973).

Mayland, H. F. and T. H. McIntosh. "Availability of Biologically-Fixed Atmospheric Nitrogen-15 to Higher Plants," *Nature* 209:421-422 (1966).

Mayland, H. F., T. H. McIntosh and W. H. Fuller. "Fixation of Isotopic Nitrogen in a Semi-Arid Soil by Algal Crust Organisms," *Soil Sci. Soc. Amer. Proc.* 30:56-60 (1966).

Rychert, R. C. "Nitrogen Fixation in Arid Western Soils," Ph.D. Dissertation, Utah State University (1975).

Rychert, R. C. and J. Skujins. "Nitrogen Fixation by Blue-Green Algae-Lichen Crusts in the Great Basin Desert," *Soil Sci. Soc. Am. Proc.* 38:768-771 (1974).

Rychert, R. C., J. Skujins, D. Sorensen and D. Porcella. "The Role of N₂ Fixation by Lichens and Free-Living Micro-Organisms in the Nitrogen Cycle of Semi-Arid to Arid Environments," in *Nitrogen Processes of Desert Ecosystems,* N. E. West and J. Skujins, eds. (Stroudsbourg, Pa.: Dowden, Hutchinson and Ross, Inc., in press).

Skujins, J. "Nitrogen Dynamics in Desert Soils," *US/IBP Desert Biome Res. Memo.* 72-40, Utah State University, Logan. (1972).

Skujins, J. "Dehydrogenase: An Indicator of Biological Activities in Arid Soils," *Bull. Ecol. Res. Comm.* (Stockholm), 17:235-241 (1973).

Skujins, J. and N. E. West. "Nitrogen Dynamics in Stands Dominated by Some Major Cold Desert Shrubs," *US/IBP Desert Biome RM* 74-42, Utah State University, Logan (1974).

Sorensen, D. L. "*In situ* Nitrogen Fixation in Cold Desert Soil-Algae Crust in Northern Utah," *M. S. Thesis,* Utah State University (1975).

Tucker, T. C. and R. L. Westerman.   "Denitrification Potential of Soils in Semi-Arid Ecosystems," *Agron. Abstr. 65th Ann. Meeting, Amer. Soc. Agron.,* Las Vegas (1973).

Westerman, R. L. and T. C. Tucker.   "Denitrification in Desert Soils," in *Nitrogen Processes of Desert Ecosystems* N. E. West and J. Skujins eds.   (Stroudsburg, Pa.:   Dowden, Hutchison and Ross, Inc., 1977).

CHAPTER 43

NITROGEN FIXATION
BY SPIRILLUM-PLANT ROOT ASSOCIATIONS

L. REYNDERS
K. VLASSAK

Katholieke Universiteit Leuven
Faculteit der landbouwwetenschappen
Laboratorium voor Bodemvruchtbaarheid en Bodembiologie
Kardinaal Mercierlaan 92
B-3030 Heverlee, Belgium

INTRODUCTION

During the last decade both increasing famine and the
rising costs of energy-consuming mineral nitrogen fertilizer
have generated soil biologists' interest in biological nitrogen fixation. Apart from the well-known *Rhizobium*-legume
symbiosis, new nitrogen-fixing associations have been dis-
covered. In tropical regions, the association between
Azotobacter paspali and the roots of *Paspalum notatum*
(Dobereiner *et al.*, 1972) and between *Spirillum lipoferum*
and the roots of *Zea mays* and different tropical grasses
(von Bülow and Döbereiner, 1975; Day and Döbereiner, 1976)
were shown to be nitrogenase active. Very recently, these
Spirillum-plant root associations were also found to be present
under temperate conditions (Okon *et al.*, 1976; Sloger and
Owens, 1976). This contribution deals with the same subject --
the occurrence and activity of nitrogen-fixing *Spirillum sp.*-
plant root associations under temperate conditions.

MATERIALS AND METHODS

Nitrogenase Activity of Bacteria-- Plant Root Associations

Plants of *Triticum aestivum*, *Zea mays* and *Cichorium
intybus* were grown under greenhouse conditions in a neutral
nitrogen-poor 2/1 (v/v) soil-sand mixture. Inoculation with
Spirillum lipoferum was performed two and three weeks after
sowing by an overhead inoculation with the bacteria suspended
in the drainage water.

During a period ranging from five to eight weeks after
sowing, nitrogenase activities of roots, rhizosphere soil and
soil away from the roots were measured using the acetylene
reduction technique. The roots were sampled late in the
afternoon, washed in distilled water and kept overnight under
an atmosphere reduced in oxygen to 2 percent by flushing with
argon. The next morning a mixture of 7.5 percent air and
10 percent C_2H_2 was injected with a disposable syringe after
complete flushing to anaerobiosis with argon. Nitrogenase
readings were performed after 8, 24 and 48 hr of incubation
in darkness at 28°C. The highest activity occurred mainly
between 8 and 24 hr after exposure to acetylene, and this
activity was used for further discussion.

Isolation of the Nitrogenase Active Organisms from the Roots

To isolate the organisms responsible for the nitrogenase
activity on the roots, 1-cm root segments were incubated in
four different semisolid media:

1. Acid medium with following composition: 20 g glucose;
   1.0 g KH_2PO_4; 0.5 g $MgSO_4$ $7H_2O$; traces $FeCl_3$ $6H_2O$;
   traces Na_2MoO_4 $2H_2O$; 1.5 g agar; 1000 ml distilled
   water.
2. Ashby's medium: 10 g glucose; 0.5 g K_2HPO_4; 0.2 g
   $MgSO_4$ $7H_2O$; 0.2 g NaCL; 0.1 g $CaSO_4$ $2H_2O$; 5 g $CaCO_3$;
   0.2 g $MnSO_4$ $4H_2O$; 0.2 mg $FeCl_3$ $6H_2O$; 0.2 mg Na_2Mo O_4
   $2H_2O$; 1.5 g agar; 1000 ml H_2O.
3. Hino and Wilson's medium (1958) containing 1.5 g agar
   to make the medium semisolid.
4. Döbereiner's nitrogen-free semisolid malate medium
   (1976).

After three days of incubation of 28°C, nitrogenase
activity of the enrichment cultures was measured using the
acetylene reduction technique. Cultures with an activity
exceeding 10 nmol C_2H_4 produced per hour were considered to
be actively involved in the nitrogen-fixing association. Great
difficulty was encountered in purifying the isolates because
in many cases the nitrogen-fixing bacteria were contaminated
by association with other organisms.

Detection of Nitrogen Fixing *Spirillum sp.* in Soil Samples

A 10% soil inoculum was incubated aerobically
in Döbereiner's medium without added bromothymol blue, and
after three days acetylene was added. After nitrogenase
activity readings, further purification was obtained by
replicate inoculation of 2-3-days old cultures into new semisolid medium up to the seventh step. The presence of
nitrogen-fixing *Spirillum sp.* was tested with the following

criteria: nitrogenase activity of the enrichment cultures and
further purified cultures, the formation of a typical *Spirillum*-
pellicle several millimeters below the medium surface, micro-
scopical examination of the pellicle-growth and physiological
tests for the hazardously obtained pure cultures.

RESULTS

In a test on the occurrence of a nitrogen-fixing
Spirillum sp. in Belgian soils, 38 samples ranging from acid
to alkaline and from sandy to clay were examined. It was
shown that 16 soils (42 percent) contained a significant
Spirillum population, 7 samples (18 percent) showed variable
results, while in 15 soil samples (40 percent) no *Spirillum*
sp. could be traced. No relation was found between soil
characteristics such as pH and texture and the occurrence
of the nitrogen-fixing organism. The samples containing
Spirillum were taken under a variable plant cover: maize,
grasses, mixed deciduous forest, barley, rye, oats, potatoes
and weed. Greenhouse experiments on *Triticum aestivum, Zea
mays* and *Cichorium intybus* were carried out to determine if
isolated *Spirillum sp.* could be nitrogenase active in the
rootsystem of temperate plants. Results on the observed
nitrogenase activity are presented in Table I.
    Apart from the observed, but not statistically signi-
ficant, differences between varieties and treatments (inoculated or not inoculated), it was observed that a severe
shortage of water in the soil could reduce N_2-ase activity
almost to zero. In one experiment, water supply to the plants
was omitted during one week, reducing the soil water content
to ± 30 percent of the WHC. In the case of wheat, no nitrogenase activity could be traced during a 48-hr exposure period,
while maize roots only regained activity after a 24-hr lag
period.
    As shown in Table II, nitrogenase activity in the plant
rhizosphere is mainly situated on the root itself. Whether
the organisms are inside the root or strongly adhered at the
root surface is not yet clear. As seen from the results, a
strong washing with distilled water could not remove the
nitrogen fixers, indicating at least a close relation between
the organism and the plant roots.
    To identify the nitrogen-fixing organism, 1-cm root
segments were incubated aerobiallly in four different media,
being more or less selective to one specific group of nitrogen
fixers. As shown in Table III, nitrogenase activity could be
found only in Döbereiner's malate medium. In the Hino and
Wilson medium, only wheat showed any activity. The organism
growing and fixing nitrogen in the malate medium was identified as a *Spirillum species*, probably identical to Beijerinck's
Spirillum lipoferum. It consists of gram-negative, very
motile small rods filled with refractile bodies, methyl red
negative, catalase positive, indole negative, not growing

Table I

Nitrogenase Activity on Roots of Different Plants   nmol C_2H_4 (g dry weight of roots)$^{-1}$hr^{-1}

Sample	Noninoculated	Inoculated	No. of Replications
Maize (mean of several varieties)	35.67	50.96	36
Maize var. Cargill Primeur 170	105.37	229.62	4
Wheat (mean of several varieties)	122.37	160.14	12
Wheat var. Clement	336.89	296.92	4
Chicory	366.05	252.35	8

Table II

Nitrogenase Activity in the Rhizosphere of Zea mays   nmol C_2H_4 (g dry weight)$^{-1}$hr^{-1}

Sample	Noninoculated	Inoculated	No. of Replications
Washed roots	34.41	92.29	12
Unwashed roots	0.29	0.49	12
Rhizosphere soil	<0.01	<0.01	12
Soil away from roots	<0.01	<0.01	12

Table III

Growth of Asymbiotic Nitrogen-Fixing Organisms Associated with Plant Roots
in Different Media as Measured by Their Nitrogenase Activity in Enrichment Cultures

Plant Species	Acid	Ashbys	Medium		Döbereiner's
			Hino and Wilson's		
Maize	-[a]	-	-		+
Wheat	-	-	+		+
Chicory	-	-	-		++++

[a]Relative scale based on following criteria:

10 nmol C_2H_4 culture^{-1} hr^{-1}

- < 10 < 50

10 < + < 50

50 < ++ < 100

100 < +++ < 200

200 < ++++

on a nitrogen-free, semisolid glucose or fructose medium and weakly growing when galactose was used as C-source. The organisms that were growing in the Hino and Wilson medium and fixing nitrogen have not been identified yet.

DISCUSSION

The results shown in Table I vary in the same range as those obtained by Döbereiner and Day (1975) working on tropical grasses:

Paspalum notatum 2-283 nmol C_2H_4/g dry weight of roots/hr

Digitaria decumbens 21-404 nmol C_2H_4/g dry weight of roots/hr

Panicum maximum 20-299 nmol C_2H_4/g dry weight of roots/hr

and those from Sylvester-Bradley (1976) working on the temperate aquatic plant *Potamogeton filiformis*, which showed mean nitrogenase activity up to 90 nmol C_2H_4/g dry weight roots/hr.

The results obtained with chicory affirm earlier findings by Vlassak and Jain (1976) who attributed a considerable nitrogenase activity to the undisturbed root-rhizosphere system of some inulin-containing plants. However, this study shows that the activity takes place on the root system more than in rhizosphere soil.

The effect of inoculation with *Spirillum lipoferum* is not identical over all the plant species tested, being positive in the case of maize but variable to negative for wheat and chicory, respectively. In the latter case the roots seem to have their own specific microbial population that can be desequilibrated by inoculation with an overdose of one single species.

The great difficulties encountered when purifying the nitrogen-fixing cultures affirm this hypothesis, for in all of the further purified isolates a characteristic companion population of contaminants was obtained, especially in the chicory isolates from which no pure *Spirillum* cultures could yet be isolated. From these preliminary experiments it can be expected that the nitrogen-fixing *Spirillum*-plant root association, perhaps in combination with other biological processes, can account for a considerable amount of biologically bound atmospheric nitrogen.

ACKNOWLEDGMENTS

The authors are grateful to I.W.O.N.L. (Instituut tot aanmoediging van het Wetenschappelijk Onderzoek in Nijverheid en Landbouw) and N.F.W.O. (Nationaal Fonds voor Wetenschappelijk Onderzoek) for financial support.

REFERENCES

Day, J. M. and J. Döbereiner. "Physiological Aspects of N_2-Fixation by a *Spirillum* from *Digitaria* Roots," *Soil Biol. Biochem.* 8:45-50 (1976).

Döbereiner, J., J. M. Day and P. J. Dart. "Nitrogenase Activity and Oxygen Sensitivity of the *Paspalum notatum-Azotobacter paspali* Association," *J. Gen. Microbiol.* 71:103-116 (1972).

Döbereiner, J. and J. M. Day. "Nitrogen Fixation in the Rhizosphere of Tropical Grasses," in *Nitrogen Fixation by Free Living Microorganisms*, W.D.P. Stewart, I.B.P. 6. (Cambridge: Cambridge Univ. Press, 1975), pp. 39-56.

Hino, S. and P. W. Wilson. "Nitrogen Fixation by a Facultative *Bacillus*," *J. Bacteriol.* 75:403-408 (1958).

Okon, Y., S. L. Albrecht and R. H. Burris. "Factors Affecting Growth and Nitrogen Fixation of *Spirillum lipoerum*," *J. Bacteriol.* 127:1248-1254 (1976).

Sloger, G. and L. D. Owens. "N_2-Fixation by a Temperate Corn-*Spirillum* Association," *II. International Symposium on N_2-Fixation*, Salamanca, Spain (1976).

Sylvester-Bradley, R. "Isolation of Acetylene-Reducing Spirilla from the Roots of *Potamogeton filiformis* from Loch Leven (Kinross)," *J. Gen. Microbiol.* 97:129-132 (1976).

Vlassak, K. and M. K. Jain. "Biological Nitrogen Fixation Studies in the Rhizosphere of *Cichorium intybus* and *Taraxacum officinale*," *Rev. Ecol. Biol. Sol.* 13:411-418 (1976).

Von Bulow, J. F. W. and J. Döbereiner. "Potential for Nitrogen Fixation in Maize Genotypes in Brasil," *Proc. Nat. Acad. Sci. U.S.A.* 72:2389-2393 (1975).

CHAPTER 44

EFFECT OF SLOW-RELEASE N-FERTILIZERS ON BIOLOGICAL N-FIXATION

L. M. J. VERSTRAETEN
K. VLASSAK

Katholieke Universiteit Leuven
Laboratory of Soil Fertility and Soil Biology
Kardinaal Mercierlaan, 92
3030 - Heverlee, Belgium

INTRODUCTION

On several occasions a rather repressive effect of mineral nitrogen on N-fixation has been demonstrated (Knowles and Denike, 1974; Mayfield and Aldworth, 1974; Rajaramamohan-Rao, 1976). Pure culture as well as soil studies have shown the effect for the nonsymbiotic N_2-fixation, whereas pot and field experiments demonstrated the same phenomenon for the symbiotic process (Hardy *et al.*, 1971).

Regarding the implications of this interaction on the basis of yield improvement and protein production, emphasis has been laid on the possible use of novel N-fertilizers (Havelka and Hardy, 1974). Furthermore, N-compounds with a controlled rate of release will preserve microbiological life and maintain the environmental conditions of the ecosystem under study (Verstraeten and Livens, 1974; Taslim and Verstraeten, 1977). Based on our experience with slow-release N-fertilizers, we investigated the effect of such nitrogen forms on the nonsymbiotic nitrogen fixation. Crotonylidene-diurea (CDU), as a member of the long-lasting N-fertilizers, and dimethylolurea (DMU), a promising urea-condensate, were tested in comparison to conventional urea by the acetylene (C_2H_2)-reduction technique. For a number of soils two improved ureaforms, UF and UFC, were also tested to establish their behavior on nitrogenase activity.

MATERIALS AND METHODS

Soils

Five soils with different characteristics were used
for this investigation (Table I). The samples were collected
from the top 0-10 cm layer of the soil profiles.

Fertilizer

To study the effect of different N-forms and their
rate of release on N_2-fixation, 10-g or 50-g portions of
soil were mixed with the N-compounds at 100 ppm of nitrogen.
The closed vials were kept at 100 percent WHC and incubated
at 30 ±1°C for varying periods of time.

N_2-Fixation

After incubation, the samples were amended with 2
percent glucose and brought into flooded conditions (Vlassak,
1973). The vials were sealed with serum caps and a needle
was inserted to equalize pressure inside and outside the
flask. After 24 hr at 30°C in the incubation room, 10 percent
of the air was replaced by acetylene with a 10-ml syringe.
Two and four hr after acetylene exposure the production of
ethylene was measured by gas chromatography. One ml of the
gas phase was taken and injected into the chromatograph
having a hydrogen flame ionization detector (FID) and a
Porapak R column. All data are the result of using four
replicate flasks, and they are represented as nmol of C_2H_2
produced per gram of dry weight soil and per hour.

RESULTS

It is quite evident from Figures 1-5 that CDU is the
only N-source that has no repressing effect on the N_2-fixing
capacity. Indeed, even at zero time, without aerobic incubation, N_2-fixation of CDU-treated samples is equal to or
even slightly better than the control samples.
DMU, as a low molecular weight urea-condensate, is
mineralized too fast, resulting in a reduced N_2-fixation for
three of the soils tested. For the other soils, degradation
was slower and hence the repressing effect of DMU was much
less or restored soon. Urea, as a fast-acting N-source,
displayed into repressive action over the entire observation
period with only a variation in the reduction ranging from
50 to 90 percent.
For a small number of soils the results of two improved
urea forms conform to their chemical characteristics, showing
a slight repressing action (Table II). Higher concentrations

Table I

Analyses of Soils

Soil	pH (H$_2$O)	Organic Carbon (%)	Total Nitrogen (%)	Sand (%) (50 p)	Clay (%) (2 p)
1	7.42	3.16	0.39	19.9	17.5
2	7.50	2.51	0.25	14.9	14.0
3	6.28	2.42	0.17	32.5	20.5
4	7.40	1.43	0.08	40.7	9.0
5	7.48	1.12	0.04	19.2	10.5

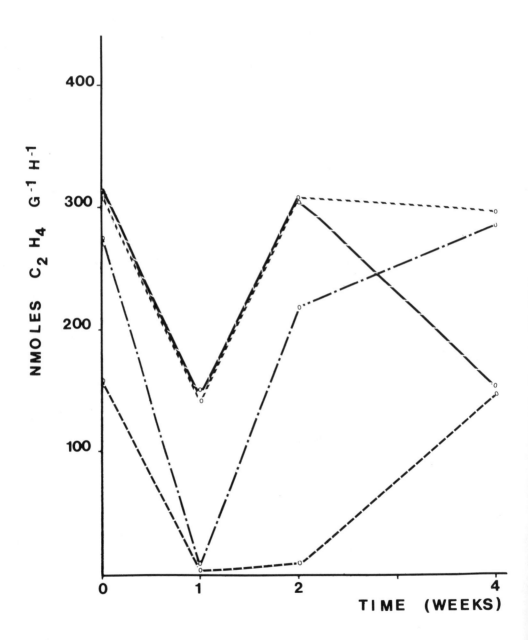

Figure 1. Rate of C_2H_2 production as a function of time for blanc ($-\;-\;-\;-$), CDU ($\longrightarrow\longrightarrow$), DMU ($\longrightarrow\cdot\longrightarrow$) and urea ($-\;-\;-\;-$).

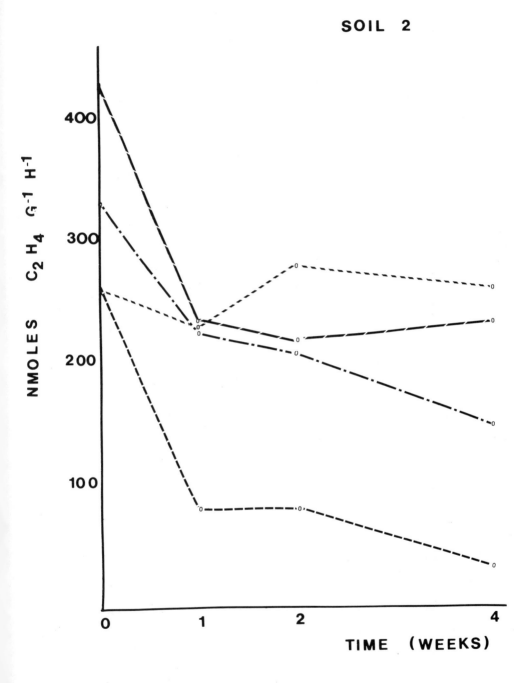

Figure 2. Rate of C_2H_2 production as a function of time for blanc (– – – –), CDU (▬▬▬▬), DMU (—·—) and urea (––––).

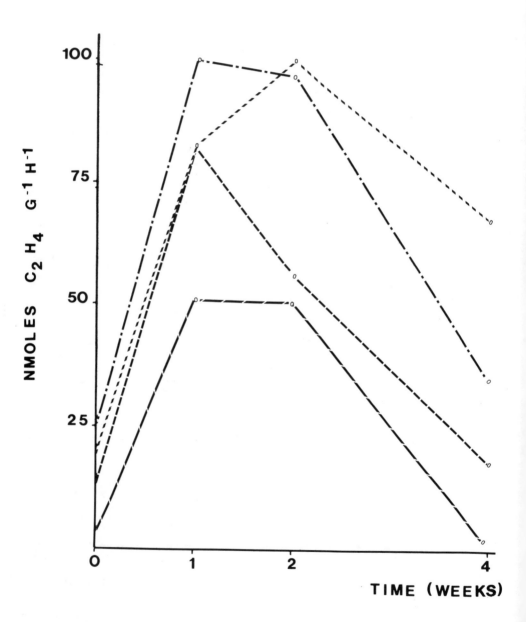

SOIL 3

Figure 3. Rate of C_2H_2 production as a function of time for blanc (‑ ‑ ‑ ‑), CDU (‑‑‑‑‑‑), DMU (‑‑·‑‑) and urea (‑‑‑‑‑).

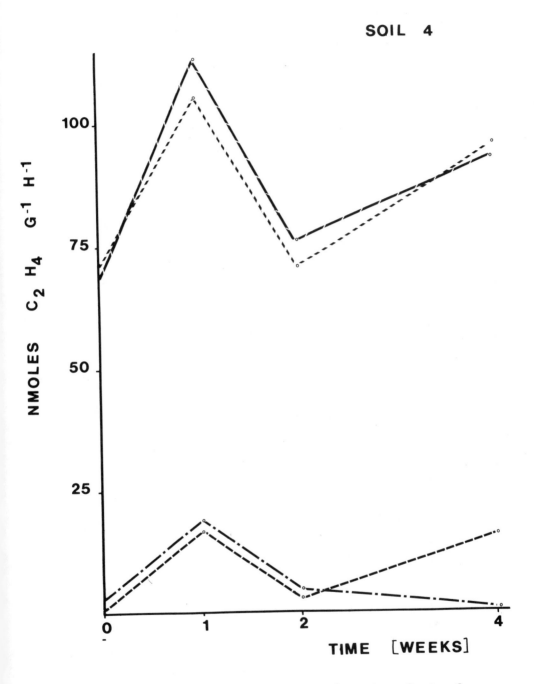

Figure 4. Rate of C_2H_2 production as a function of time for blanc (– – – –), CDU (▬▬▬▬), DMU (▬·▬) and urea (– – – –).

SOIL   5

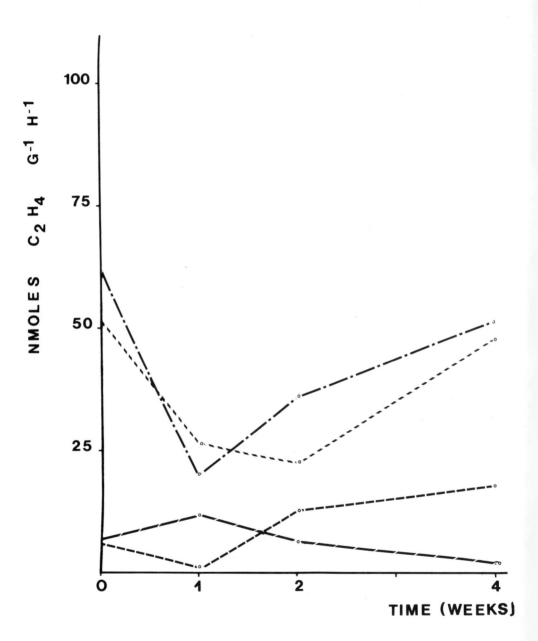

Figure 5.   Rate of C_2H_2 production as a function of time for blanc (– – – –), CDU (▬▬▬), DMU (▬ · ▬) and urea (▬ ▬ ▬).

Table II

Effect of UF and UFC on the C_2H_2-Reducing Activity in Waterlogged Soil Amended with Glucose

Treatment	Mean Values nmol $C_2H_2/g^{-1}/h^{-1}$ Incubation (days)			
	0	7	14	28
Soil 2	263.8	229.7	279.0	260.5
+ UF	266.1	225.8	114.7	99.6
+ UFC	152.4	142.7	115.0	69.7
Soil 3	19.7	82.7	103.1	
+ UF	22.3	146.0	104.7	
+ UFC	16.0	121.0	108.9	
Soil 4	71.4	106.6	71.9	95.5
+ UF	53.9	56.5	45.7	57.4
+ UFC	76.4	88.8	61.2	72.7
Soil 5	52.1	26.6	22.7	
+ UF	39.0	47.0	45.0	
+ UFC	40.2	35.0	47.0	

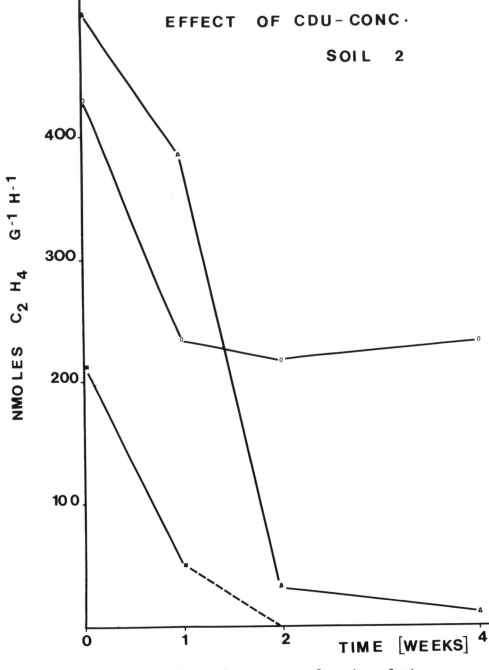

Figure 6.   Rate of C_2H_2 production as a function of time
   for different concentrations of CDU: 100 ppm (■———■),
   200 ppm (△———△), 1000 ppm (○———○).

of CDU, in fact, showed that N_2-fixation becomes involved due to the amount of mineral nitrogen present in the soil medium (Figure 6). At 100 ppm concentration, mineralization experiments proved that only 15 to 20 percent of CDU is available within the first week.

DISCUSSION

The data presented in this paper (Figures 1-6 and Table II) show that the process of N_2-fixation is greatly affected by the available nitrogen, which in turn is governed by the N-form applied. In spite of fluctuations in time, the nitrogenase activity remains fairly normal over the entire observation period. So far, only crotonylidenediurea (CDU), a commercial slow-release N-fertilizer, is able to preserve the N_2-fixing capacity. This phenomenon is quite interesting in terms of fertilization of symbiotic C-fixing crops such as soybeans and other legumes to improve the yield beyond the theoretical limit, as mentioned by Sinclair and De Wit (1975, 1976). In general, such products might be interesting tools for a more biological approach to the environmental problems of leaching, waste and pollution. Nitrogen forms with a rate or release comparable to the native soil-mineralization will become more and more important to man.

The effect of DMU fluctuates much more, depending on its rate of release, which is affected by the soil itself. Therefore, if DMU-degradation is less than 80 percent of the urea mineralization, it displays a less repressive action or at least it is faster in restoring N_2-fixation.

The urea forms, UF and UFC, give more reproducible results similar in nature to CDU, but their rate of release is slightly faster and their repressing effect a bit more pronounced. The careful investigation of a large number of promising slow-release N-fertilizers, by biological tests such as N_2-fixation, will enable us to evaluate those products for new developments in fertilization and environmental policy.

ACKNOWLEDGMENTS

The authors express their gratitude to the I.W.O.N.L. (Instituut tot aanmoediging van het Wetenschappelijk Onderzoek in Nijverheid en Landbouw) and to the N.F.W.O. (National Fonds voor Wetenschappelijk Onderzoek) for the financial support. In addition the following companies are also acknowledged for their kind provision of N-fertilizers: B.A.S.F., Germany (CDU), Fertilizers and Chemicals Ltd., Israel (UFC), and Mitsui Toatsu Chemicals Ltd., Japan (UF).

REFERENCES

Hardy, R. W. F., R. C. Burns, R. R. Hebert, R. D. Holsten and E. K. Jackson. "Biological Nitrogen Fixation: A Key to World Protein," *Plant and Soil* Spec. vol.:561–590 (1971).

Havelka, U. D. and R. W. F. Hardy. "Legume N_2-Fixation as a Problem in Carbon Nutrition," Intern. Symp. on Nitrogen Fixation at Pullman, Washington, June 3–7, 1974, W. E. Newton and C. Y. Nyman, eds. (Pullman, Washington: Washington State University Press, 1974), pp. 1–30.

Knowles, R. and D. Denike. "Effect of Ammoniumnitrite and Nitrate-Nitrogen on Anaerobic Nitrogenase Activity in Soil," *Soil Biol. Biochem.* 6:353–357 (1974).

Mayfield, C. I. and R. L. Aldworth. "Acetylene Reduction by Nonsymbiotic Bacteria in Artificial Soil-Aggregates Amended with Glucose," *Can. J. Microbiol.* 20:877–881 (1974).

Rajaramamohan-Rao, V. "Nitrogen Fixation as Influenced by Moisture Content, Ammonium Sulphate and Organic Sources in a Paddy Soil," *Soil Biol. Biochem.* 8:445–448 (1976).

Sinclair, T. R. and C. T. De Wit. "Photosynthethate and Nitrogen Requirements for Seed Production by Various Crops," *Science* 189:565–567 (1975).

Sinclair, T. R. and C. T. De Wit. "Analysis of the Carbon and Nitrogen Limitations to Soybeans Yield," *Agron. J.* 68:319–324 (1976).

Taslim, H. and L. M. J. Verstraeten. "Mineralization of Slow-Release N-Compounds in Water-logged Conditions," *Z. Pflanzenern. Bodenk.* (in press).

CHAPTER 45

TEMPERATURE DEPENDENCE OF NITRATE LOSSES AND DENITRIFIER POPULATIONS IN SOIL COLUMNS

H. E. DONER
A. D. MCLAREN

Department of Soils and Plant Nutrition
University of California
Berkeley, California 94720 USA

INTRODUCTION

Biological denitrification is known to be temperature-dependent (Nommick, 1956; Bremner and Shaw, 1958; Cooper and Smith, 1963; Bailey and Beauchamp, 1973; and Stanford *et al.*, 1975); rates increase with temperature from 3 to 65° C (Nommick, 1956). Above 65° C denitrification ceases, presumable due to inhibition and/or inactivation of the denitrifiers. Thus far, few have tried to relate denitrification rates to denitrifier populations. It is clear, however, that an understanding of transformation rates in biogeochemical systems requires an assessment of the magnitude of the pertinent biomass (McLaren, 1973).

Most studies concerned with denitrification rates have been conducted in closed, culture flask-type systems. In these systems the concentrations of NO_3^- and "available" organic matter decrease with time. Ostensibly, the denitrifier population may first increase and then decline in an anoxic environment (or microenvironment). Soil columns provide a method of studying denitrification in which a constant rate of addition of nitrate and growth of the denitrifiers can be monitored (Doner *et al.*, 1975).

The purpose of this study was to determine the effect of temperature on rates of denitrification and changes of the denitrifier populations. Temperature ranged from 15 to 35° C, as can happen in a field.

METHODS AND MATERIALS

The experiment was conducted by packing glass columns (2.6 cm i.d.) with a 2-mm particles of a Hanford sandy loam (Typic xerorthent) to a height of 10 cm giving a bulk density of 1.71 g/cm^{-3}. The temperature of the columns was kept relatively constant (±0.1° C) with the aid of water jackets. The temperatures studied were 15, 25 and 35° C. Solutions containing 1 me ml^{-1} CaCl$_2$ and 0, 25, 50 and 100 µg ml^{-1} NO$_3^-$-N were applied with a peristaltic pump at the bottom and collected at the top of the vertical columns. Oxygen entry was possible only as dissolved gas in the influent solution. Two columns with 25 and 100 µg ml^{-1} NO$_3^-$-N and one column with zero and one with 50 µg ml^{-1} NO$_3^-$-N were infiltrated for a total of six columns.

At the end of seven days, one each of the two 25- and 100-µg ml^{-1}NOO$_3^-$-N treatments was terminated. The remaining three treatments were continued until the tenth day. Other trials at room temperature showed that after seven days the NO$_3^-$ concentration in the effluent increased very slowly. Effluent NO$_3^-$ and NO$_2^-$ concentrations were determined colorimetrically every 12 hr. Soil was removed from columns at the termination of the experiments and split into top and bottom 5-cm sections (approximate) and samples were taken for microbiological counts. Denitrifier counts (Alexander, 1965a) and total cell counts (Clark, 1965) were determined by a most probable number method (Alexander, 1965b).

RESULTS AND DISCUSSION

Increasing the temperature from 15 to 35° C increased NO$_3^-$ loss in any treatment (Table I). Days 7.5 through 10 were used to estimate NO$_3^-$ losses. Changes in NO$_3^-$ losses and flow rates were small during this interval, probably because of biological denitrification since assimilation would not account for significant amounts of nitrogen utilization (Doner *et al.*, 1974). A small amount of NO$_3^-$ was observed in the effluents with 0 treatment. It is not known whether this represents a small production of NO$_3^-$ or some interference from the soil solution in the NO$_3^-$ analysis.

Traces of NO$_2^-$ were observed in all effluents; an increase in temperature increased the concentration of NO$_2^-$ in the effluents. No apparent relationship existed between effluent NO$_3^-$ and NO$_2^-$ concentrations. Maximum NO$_2^-$ concentrations, of 5 to 30 µg/ml^{-1} N, were found in the first leachates from the soil columns at 25 and 35° C. At 15° C, maximum NO$_2^-$ concentrations of about 3 µg/ml^{-1} N were observed 18 to 24 hr after the first leachates were collected. This peak NO$_2^-$ concentration corresponded well with the beginning of NO$_3^-$ losses. It appears that a delay of 18 to 24 hr at 15° C was

Table I
The Effect of Temperature and NO_3^- Concentration
in the Influent or Effluent NO_3^- and NO_2^- Concentrations

Temperature	Influent NO_3^- Conc. ($\mu g\,N\,ml^{-1}$)	Effluent NO_3^- Conc. ($\mu g\,N\,ml^{-1}$) Last 3-Day	Effluent NO_2^- Conc. ($\mu g\,N\,ml^{-1}$) Average
	0	0.18 ± 0.01	0.017
	25	19.6 ± 0.6	0.040
15°	50	45.3 ± 0.7	0.058
	100	92.7 ± 1.0	0.055
	0	0.21 ± 0.03	0.029
25°	25	16.5 ± 1.7	0.084
	50	40.0 ± 2.4	0.10
	100	95.0 ± 1.5	0.083
	0	0.12 ± 0.03	0.013
35°	25	7.6 ± 2.6	0.14
	50	34.2 ± 1.9	0.26
	100	81.8 ± 1.8	0.16

necessary for the microbial population to cause denitrification. This may have been due to slow microbial growth, reduced metabolic demands, and slower creation of anaerobic conditions.

By knowing the amount of NO_3^- lost, flow rate (average 0.45 cm/hr^{-1}; range 0.30 to 0.60 cm/hr^{-1}), percent saturation (75%), and soil column length, denitrification rates can be estimated for each treatment. The temperature dependence of rates in terms of Arrhenius' relationship is

$$k = k_\infty \exp(-E_a/RT) \qquad (1)$$

A plot of log denitrification rates (k) against reciprocal of absolute temperature based on Equation 1 is shown in Figure 1. The frequency factor k_∞ is equal to the

$$\left(\lim_{T \to \infty}\right)k.$$

E_a is activation energy in cal mole^{-1}; R is the ideal gas constant, 1.987 cal mole^{-1} °K^{-1} and T is the temperature in degrees Kelvin.

Figure 1 shows denitrification rates as a function of temperature based both on grams of soil, k_s, and on denitrifier populations at the effluent end of columns (Figure 2), k_b. The empirical equation for k_s is:

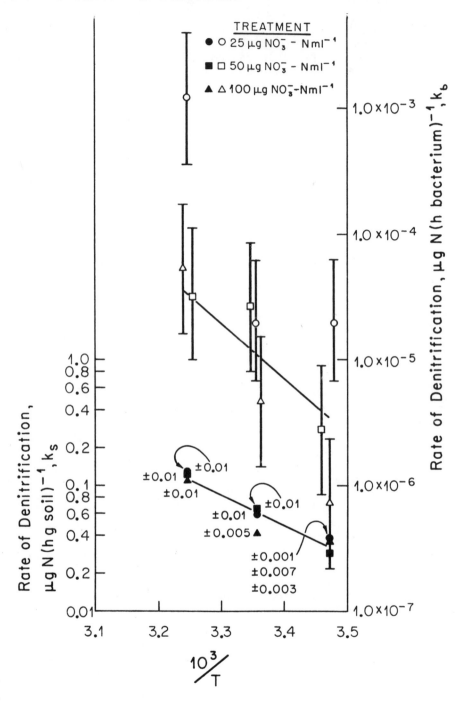

Figure 1.   Effect of temperature on zero order denitrification rates based on grams of soil (k_s) and populations of denitrifiers (k_b). Shaded symbols are for k_b and open symbols for k_s. Vertical lines indicate 95 percent confidence limits for rates of denitrification.

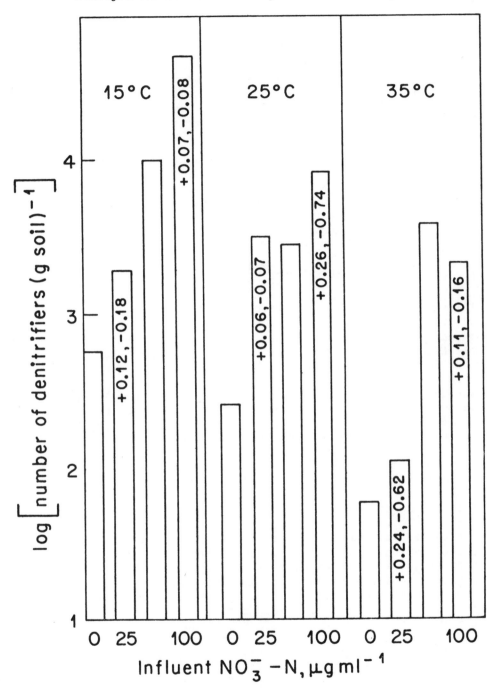

Figure 2. Log of denitrifier population in effluent end-half of soil column. Twenty-five and 100-μg NO$_3^-$-N ml^{-1} values are averages of seven and ten day counts.

$$\log k_s = 7.05 - 2.46 \left(\frac{10^3}{T}\right) \qquad (2)$$

with an activation energy, E_a, of 11 kcal mol^{-1}. The empirical equation for k_b is:

$$\log k_b = 9.69 - 4.36 \left(\frac{10^3}{T}\right) \qquad (3)$$

and in this case E_a = 20 kcal mol^{-1}. Activation energies of 10 to 20 kcal mol^{-1} are common with biological systems (Bray and White, 1966, p. 166). Thus E_a based on soil mass lies at the lower limits of the range and E_a based on denitrifier population at the upper limits.

The Q_{10} (temperature coefficient) for denitrification in the soil columns was 1.9 between 15 and 25° C where

$$Q_{10} = \frac{k_s \text{ at } T + 10°}{k_s \text{ at } T} \qquad (4)$$

This Q_{10} value agrees well with those reported by Stanford *et al.*, (1975) and Cooper and Smith (1963). The former investigators used first order kinetics to describe the denitrification reaction. Substituting values of k_b for k_s in Equation 4 gives a Q_{10} of 3.2, a substantially greater temperature effect. Although uncertainties in denitrifier counts and low coefficient of linear correlation ($r = 0.42$) for k_b necessitate obtaining more precise data before an interpretation of these differences in Q_{10} values can be made, it is clear that rates on a per gram of soil basis are not equivalent to rates on a per microbe per gram of soil basis.

Denitrifier counts as related to NO_3^- treatment and temperature, from soil in the column half closest to the effluent end, are in Figure 2. Denitrifier populations in the soil in the column half closest to the solution entrance were consistently higher than at the other end, as shown by the ratios in Table II. This probably resulted from a higher concentration of NO_3^- at the influent end of the soil. Effluent ends of the soil columns are thought to have been more anoxic (incoming solutions contained dissolved air). Two observations are gleaned from Figure 2. First, the addition of NO_3^- in leaching solutions increased denitrifier population over the check treatments (0 µg/ml^{-1} NO_3^--N) as expected. Second, a *general* decrease in denitrifier population with increased temperature for any given NO_3^- treatment probably is manifested. In an earlier report Doner *et al.* (1975) found a denitrifier population of 10^4 cells g^{-1} soil after 48 hr at room temperature and with 100 µg NO_3^--N ml^{-1} in influent solution. That compares with the population found in this study after seven to ten days. With lower NO_3^- concentration and lower available organic matter, the denitrifier population was reduced.

Table II

Ratios of Denitrifier Populations at the Influent Half of Soil Columns (I)
to Effluent Half of Soil Columns (E) and Average Total Cell Counts in Whole Column

Temperature	NO$_3$ Treatment (μg N ml^{-1})	Ratio of Denitrifier Population (I/E)	Average Total Cell Count (cells g^{-1} soil x 10^{-7})
15°	0	68	0.58 ± .19
	25	72	0.98 ± 0.01
	50	10	1.5 ± 0.0
	100	3	1.8 ± 0.3
25°	0	1500	0.65 ± 0.35
	25	47	1.3 ± 1.1
	50	14	1.4 ± 0.4
	100	10	1.2 ± 0.2
35°	0	260	0.30 ± 0.03
	25	1400	0.46 ± 0.08
	50	8	0.57 ± 0.09
	100	3	0.69 ± 0.14

The average denitrifier counts found for 25 and 50 µg NO_3^--N ml^{-1} treatments (Figure 2) were averages of 7 and 10 day samples. In most cases the seven-day populations were higher than those at ten days (except for 100 µg NO_3^--N ml^{-1} at 35° C). This trend, if real, may be due to reductions of available organic matter with time. It appears that either the efficiency of the denitrifier population was greater at higher temperatures or the method of enumerating the populations was not sensitive to soil conditions, since higher temperatures increased the rates of denitrification without a corresponding increase in denitrifiers.

In treatments in which no NO_3^- was added, denitrifiers were also found. This may reflect indigenous nitrate and a resting state. Higher populations at lower temperatures may be noted as a trend, which means that the influence of temperature on growth of denitrifiers and rates of denitrification are opposite. Even the control columns show a decrease of populations with an increase of temperature, so at a higher temperature organic matter consumption may lead to starvation. Total cell counts were about the same at 15 and 25° C independent of NO_3^- treatments (Table II). At 35° C lower *total* populations were found just as with the denitrifiers.

SUMMARY

In this study rates of denitrification and denitrifier populations were determined at 0, 25, 50 and 100 µg NO_3^--N ml^{-1} solution influents in soil columns at 15, 25 and 35° C. Increased temperatures decreased denitrifier populations with any given treatment. Rates of denitrification based on weight of soil increased by a factor of 1.9 by increasing the temperature from 15° to 25° C; whereas it increased by a factor of 3.2 based on denitrifier population. There is a need to determine available organic matter also in order to help evaluate denitrifier efficiency.

ACKNOWLEDGMENTS

Appreciation is extended to Dr. L. W. Belser for making microbial counts and Mr. A. Pukite for the chemical analyses. This work was supported partially by RANN-NSF Grant G143664 and the Kearney Foundation of Soil Science.

REFERENCES

Alexander, A. "Denitrifying Bacteria," in *Methods of Soil Analysis. Part 2. Chemical and Microbiological Properties*, C. A. Black, ed., Agron. Series Number 9 (Madison, Wisconsin: American Society of Agronomy, 1965a), pp. 1484-1486.

Alexander, M. "Most-Probably-Number Method for Microbial Count," in *Methods of Soil Analysis. Part 2. Chemical and Microbiological Properties*, C. A. Black, ed., Agron. Series Number 9 (Madison, Wisconsin: American Society of Agronomy 1965b), pp. 1467-1472.

Bailey, L. D. and E. G. Beauchamp. "Effects of Temperature on Nitrate and Nitrite Reduction, Nitrogenous Gas Production and Redox Potential in a Saturated Soil," *Can. J. Soil Sci.* 53:213-218 (1973).

Bray, Geoffrey H. and Kenneth White. *Kinetics and Thermodynamics in Biochemistry*. 2nd ed. (New York/London: Academic Press Inc., 1966).

Bremner, J. M. and K. Shaw. "Denitrification in Soil. II. Factors Affecting Denitrification," *J. Agr. Sci.* 51:40-52 (1958).

Clark, F. E. "Agar-plate Method for Total Microbial Count," in *Methods of Soil Analysis. Part 2. Chemical and Microbiological Properties*, C. A. Black, ed., Agron. Series Number 9 (Madison, Wisconsin: American Society of Agronomy 1965), pp. 1460-1466.

Cooper, G. S. and R. L. Smith. "Sequence of Products Formed During Denitrification in Some Diverse Western Soils," *Soil Sci. Soc. Am. Proc.* 27:659-662 (1963).

Doner, H. E., M. G. Volz and A. D. McLaren. "Column Studies of Denitrification in Soil," *Soil Biol. Biochem.* 6:341-346 (1974).

Doner, H. E., M. G. Volz, L. W. Belser and Jan-Per Loken. "Short-Term Nitrate Losses and Associated Microbial Populations in Soil Columns," *Soil Biol. Biochem.* 7:261-263 (1975).

McLaren, A. D. "A Need for Counting Microorganisms in Soil Mineral Cycles," *Environmental Letters* 5:143-154 (1973).

Nommik, H.   "Investigation on Denitrification in Soil,"
  Acta Agric. Scand. 6:195-228 (1956).

Stanford, G., S. Dzienia and R. A. Vander Pol.   "Effect of
  Temperature on Denitrification Rate in Soils," *Soil Sci.
  Soc. Am. Proc.* 39:867-870 (1975).

CHAPTER 46

NITRIFICATION RATE IN A CONTINUOUS FLOW SYSTEM

M. J. BAZIN
DIMITY J. COX

Department of Microbiology
Queen Elizabeth College
Campden Hill Road
London W.8., England

INTRODUCTION

Investigation of nitrification dynamics in soil using the continuous flow technique of Macura and Malek (1958) was undertaken first by Macura and Kunc (1965). McLaren and his co-workers also have used this method and combined it with an analysis of the system in terms of mathematical models (see Doner and McLaren, 1976). This combination of a theoretical and experimental approach has been employed in our laboratory, where we have attempted to minimize the physicochemical effects of soil in order to isolate the biological characteristics of the system by using glass beads instead of soil as the packing material in our columns (Bazin and Saunders, 1973). Although this approach lessens the direct applicability of experimental results to field situations, we believe that we are more likely to reveal significant biological processes by eliminating as many undefined environmental variables as possible.

Two fundamental properties of nitrification columns are the downward flow of nutrients and products and the adhesion of the nitrifiers to the packing material. In this chapter we attempt to use this latter property to interpret the results of experiments in which the nitrification rate in columns containing *Nitrosomonas europaea* and *Nitrobacter agilis* was measured.

MATERIALS AND METHODS

The methods employed were essentially those of Bazin and Saunders (1973) except where indicated. Nitrification rate was measured in terms of the capacity of the whole column to convert ammonium to nitrite and nitrate. It was estimated as the product of the flow rate and the concentration of these ions in the effluent solution. The effect of altering two environmental parameters, flow rate and input ammonium concentration, on the nitrification rate was tested.

RESULTS

Figure 1 shows how the steady-state nitrification rate changed as a function of flow rate. Within the range of flow rates employed, the response appeared to be linear. The way

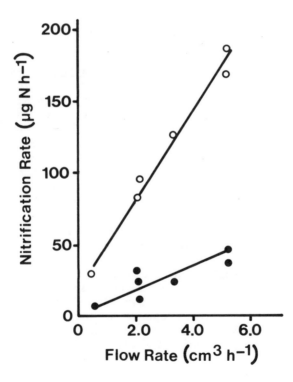

Figure 1. Nitrification rates at steady state as a function of flow rate in a column packed with glass beads of average diameter 0.18 mm and supplied with 106 µg cm^{-3} NH_4^+-N. Open circles NO_2^--N; closed circles NO_3^--N.

in which nitrification rate changed after the flow rate was altered depended on whether the flow rate was increased or decreased. An increase in flow rate caused a sudden increase in the nitrification rate and then a decrease to the new steady state as illustrated in Figure 2. A decrease in flow

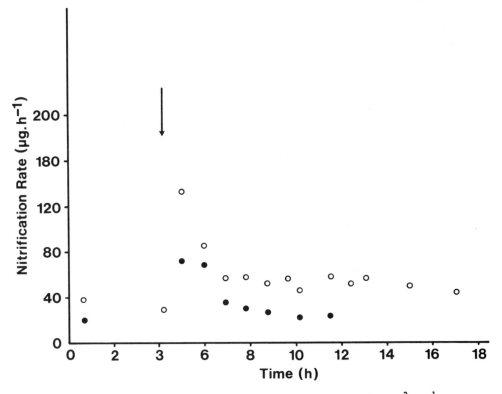

Figure 2. Effect of changing flow rate from 1.41 cm^3hr^{-1} to 3.55 cm^3hr^{-1} at time indicated by the arrow. Symbols and column specifications as in Figure 1.

rate resulted in an analogous decrease in nitrification rate and, in addition, an overshoot prior to the new steady state. An example of a shift-down in flow rate is shown in Figure 3.

Figure 4 illustrates the way nitrification rate at steady-state changed as a function of the concentration of ammonium supplied to a nitrification column. Up to about 90 µg cm^{-3} NH_4^+-N the relationship appears linear but after this point some factor other than ammonium seems to have been limiting. As can be seen in Figure 4 all the ammonium that oxidized was converted to nitrate except at the steady state, in which 10 µg cm^{-3} NH_4^+-N was supplied. In this case most of the ammonium was converted to nitrite. This apparently anomalous result occurred before a higher concentration of ammonium was supplied to the column.

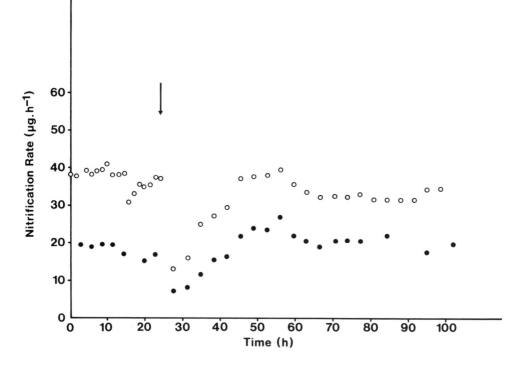

Figure 3.   Effect of changing flow rate from 3.53 cm^3hr^{-1} to 1.46 cm^3hr^{-1} at time indicated by the arrow.   Symbols and column specifications as in Figure 1.

Figure 5 shows what happened when the ammonium concentration was increased from 10 μg cm^{-3} to 63.5 μg cm^{-3} and then decreased back to 12 μg cm^{-3} NH_4^+-N.   Although prior to this (approximately) square wave perturbation only about 10 percent of the ammonium was oxidized to nitrate, afterwards 100 percent conversion occurred.

DISCUSSION

Chi, Howell and Pawlowsky (1974) have shown both theoretically and experimentally that multiple steady states are possible in chemostat cultures in which wall growth occurs and substrate inhibition is taking place.   Adhesion of nitrifiers to a solid substratum is likely, therefore, to be responsible for the occurrence of initial conditions-dependent steady states in nitrification columns, particularly as nitrification kinetics are subject to inhibition by the substrates and products of the process.

The behavior of the nitrification column when the flow rate was changed is less readily explained in terms of the habitat of the nitrifiers although the asymmetric response

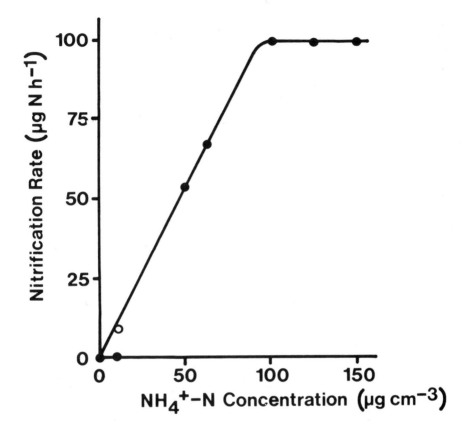

Figure 4. Nitrification rate at steady state as a function of concentration of ammonium in the input medium. The glass beads had an average diameter of 1.07 mm, and the flow rate through the column was maintained at 2.15 cm^3hr^{-1}. Symbols as in Figure 1.

we observed does not seem to occur in chemostat cultures (Scott, personal communication) nor is it expected on theoretical grounds when Monod (1942) saturation kinetics are used to model a well-mixed system. When such kinetics are used to model a column, however, asymmetric overshoots in effluent ion concentrations can occur (Prosser, 1976).

ACKNOWLEDGMENTS

    This research was supported by an SRC Research Grant and an SRC CASE award sponsored by the Water Research Centre.

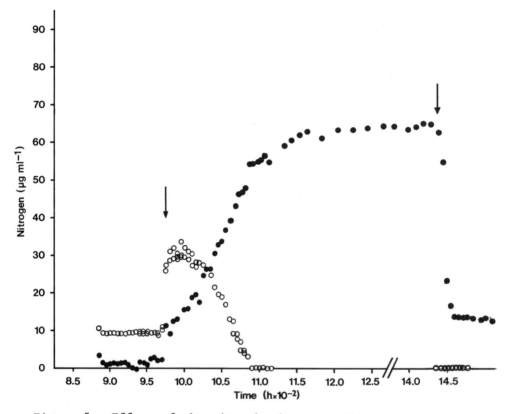

Figure 5.   Effect of changing the input nutrient concentration. At the arrows the ammonium concentration was changed from 10 μg cm^{-3} to 63.5 μg cm^{-3} and then to 12 μg cm^{-3} NH$_4^+$-N. Symbols and column specifications as in Figure 4.

REFERENCES

Bazin, M. J. and P. T. Saunders. "Dynamics of Nitrification in a Continuous Flow System," *Soil. Biol. Biochem.* 5:531–543 (1973).

Chi, C. T., J. A. Howell and V. Pawlowsky. "The Regions of Multiple Stable Steady States of a Biological Reactor with Wall Growth, Utilizing Inhibitory Substrates," *Chem. Eng. Sci.* 29:207–211 (1974).

Doner, H. E. and A. D. McLaren. "Soil Nitrogen Transformations: A Modelling Study," in *Environmental Biogeochemistry.* Vol. 1, J. O. Nriagu, ed. (Ann Arbor, Michigan: Ann Arbor Science Inc., 1976), pp. 245–258.

Macura, J. and F. Kunc. "Continuous Flow Method in Soil Microbiology. V. Nitrification," *Folia Microbiol.* Praha, 10:125–135 (1965).

Macura, J. and I. Malek. "Continuous-Flow Method for the Study of Microbiological Processes in Soil Samples," *Nature* 182:1796 (1958).

Monod, J. *Recherches sur la Croissance des Cultures Bacteriennes* (Paris: Hermann et Cie, 1942).

Prosser, J. I. "A Model System for Nitrification," Ph.D. Thesis, University of Liverpool, England (1976).

CHAPTER 47

CALCULATIONS OF THE NITRITE DECOMPOSITION REACTIONS IN SOILS

O. VAN CLEEMPUT
L. BAERT

Faculty of Agriculture
University of Ghent
Ghent, Belgium

INTRODUCTION

Although nitrite usually does not accumulate in high concentrations in soils or terrestrial environments, it may appear during several nitrogen transformation processes.

Waterlogging a soil results in rapid depletion of oxygen and successive reduction of nitrate and other oxidized soil components. The first step in the reduction of nitrate is the formation of nitrite. Besides being produced as an intermediary component in denitrification, nitrite can also be formed during nitrification of ammonium or ammonium-yielding fertilizers. The accumulation of nitrite depends on different factors such as ammonium level, pH (Chapman and Liebig, 1952; Tyler and Broadbent, 1960), organic matter (Smith, 1964), temperature and moisture (Justice and Smith, 1962) and soil-fertilizer geometry (Bezdicek *et al.*, 1971).

Chemical decomposition of nitrite may be one of the processes responsible for loss of nitrogen out of the soil. Important work has been done on some factors influencing this nitrite decomposition (Nelson and Bremner, 1969, 1970; Bulla, Gilmour and Bollen, 1970; Van Cleemput *et al.*, 1976; Wullstein and Gilmour, 1964, 1966). Soil pH, organic matter and metallic cations seem to be the key factors.

The major gases evolved during chemical nitrite decomposition are, according to Bollag *et al.*, (1973), nitric oxide and nitrogen dioxide. Nelson and Bremner (1970) found that substantial amounts of N_2 and NO_2 and small amounts of N_2O were evolved on treatment of neutral and acidic soils with nitrite. About the different decomposition products and their relative importance, the literature data are quite divergent.

Therefore, it is the aim of the authors to calculate the spontaneity of different nitrite decomposition reactions and to obtain information on the most important decomposition products that may be formed spontaneously in nonstandard state conditions. This is in contrast to some calculations worked out by Chao and Kroontje (1963) and related to standard-state conditions.

This chapter will try to answer the following questions:

- In what environmental conditions is nitrite chemically stable or unstable?

- What kind of decomposition products can be formed spontaneously and how is their relative stability?

- What is the influence of the environmental factors O_2 and pH on the number and abundance of intermediary compounds?

It should be pointed out that this chapter is limited to those nitrite decomposition reactions whereby nitrite is transformed to another nitrogen compound, excluding the self-decomposition reactions. Calculations on some of these reactions were the subject of another paper (Van Cleemput and Baert, 1976).

PROCEDURE

The mode of calculation and subsequent interpretation of the results are based on the use of the Gibbs free reaction energy (ΔG_r). The sign of this value indicates in what direction the equilibrium is attained. A negative ΔG_r value indicates that equilibrium is situated to the right so that spontaneous change occurs.

The nitrite decomposition reactions were selected from a previously published list (Van Cleemput and Baert, 1974) of nitrogen reactions. Involved are those reactions whereby nitrite is the only nitrogen reactant and the following N-compounds as reaction products: N_2, N_3H, NH_4^+, N_2H_2, NO $N_2H_5^+$, N_2O, $H_2N_2O_2$, N_2O_3, NH_2, NH_3OH^+, NO_2, NO_3, N_2O_4, N_2O_5, and HNO. Taking into account the Gibbs standard free formation energy (ΔG_f^o) of the different participating reaction components, the Gibbs standard free reaction energy is calculated. For the N-compounds having protolytic properties, changes of their species distribution due to pH variations having an influence on the calculation of ΔG_r^o and ΔG_r were taken into consideration according to the procedure earlier described (Van Cleemput and Baert, 1974).

The ΔG_r values were calculated for each of the 16 reactions, taking into account the respective ΔG_r^o values and the following set of variables:

- activity of nitrite and the participating nitrogen compounds varying from 10^{-2} to 10^{-20};

- pH values: 4, 7 and 9;

- O_2 activities: 0.2, 10^{-2}, 10^{-4}, 10^{-8}, 10^{-16}.

If all considered reactions have the opportunity to occur simultaneously, the reaction with the lowest negative ΔG_r should be the one continuing to equilibrium. This means that the formed product arising from this reaction will be the main product (most stable) at equilibrium. By calculating simultaneously the ΔG_r value of all reactions at a certain O_2 activity, pH and a certain set of activities of nitrite and the corresponding N-product, the ΔG_r values can be compared and the lowest negative value will be picked out. That particular reaction will proceed dominantly to the right. It also means that the formed product(s) is(are) the main product(s) (Most stable) at equilibrium. Removing that particular reaction from the list of reactions and repeating the same calculation procedure allows establishment of a chain of decomposition products of decreasing stability relative to their particular activity, pH and O_2 activity.

RESULTS AND DISCUSSION

The results of the calculations of the ΔG_r values as a function of pH, oxygen activity and activities of the participating reaction components for the most stable product are presented in Figure 1. The three cubes represent the results at pH 4, 7 and 9. The axes represent the activities of nitrite, O_2 and the reaction product, respectively. Each cube is divided into a lower part and an upper part. The lower part represents the area wherein NO_2^- is stable. This means a positive ΔG_r value for all considered reactions. The upper part represents the area wherein NO_2^- is not stable and some reactions may have negative ΔG_r values. From Figure 1 it can be seen that the area of nitrite stability (upper part) is mostly filled by N_2. Only at very low product activity and aerobic conditions does NO_2 take over the dominance, irrespective of pH. The stability area of nitrite (lower part) increases by increasing the pH from 4 to 9. The lower the O_2 tension and product activity, the easier NO_2^- decomposes. These theoretical results coincide with experiments in soils and solutions showing that the stability of nitrite decreases by decreasing pH and O_2 tension (anaerobic conditions).

By removing from the list of reactions the reaction resulting in N_2 production, another reaction will have the lowest ΔG_r value. In this case the nitrite instability area is filled up by nitrous oxide and nitrogen dioxide.

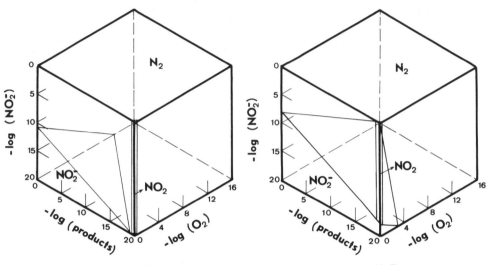

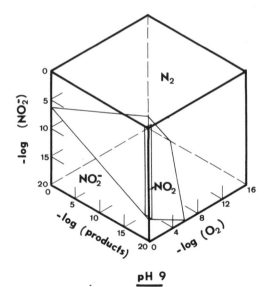

Figure 1.   Representation of the regions of stable reaction
   products at equilibrium, taking into account all 16 reactions.

Simultaneously, the nitrite stability area is extended
(Figure 2). It can be seen that N_2O becomes more important
relative to NO_2, at lower O_2 activities or in anaerobic media.
The subsequent step besides removing the N_2-producing reaction,
was also to avoid the reaction producing NO_2. The results of
this are presented in Figure 3. The area of nitrite instability
is divided up by N_2O and NO. It can be seen that the ratio
of the N_2O area to the NO area remains constant from pH 4 to
9. Both nitrogen compounds, however, have a smaller stability

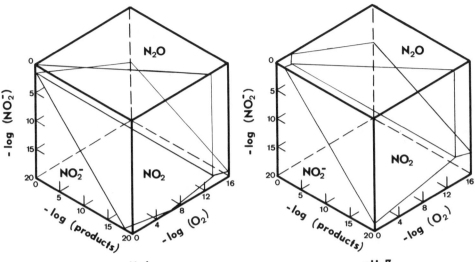

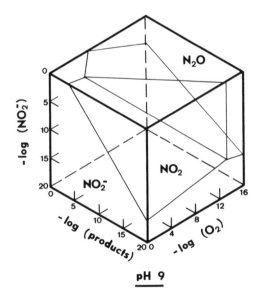

Figure 2. Representation of the regions of stable reaction products taking into account all reactions except the N_2-producing one.

area in alkaline conditions. Experiments on nitrite reduction in soils of different acidity also show less nitrogen oxide compounds in alkaline conditions.

Continuing the procedure of subsequent avoidance of the reaction forming the most stable product, one can establish a list of nitrogen products of decreasing stability. This list spontaneously formed out of NO_2^- as a function of O_2

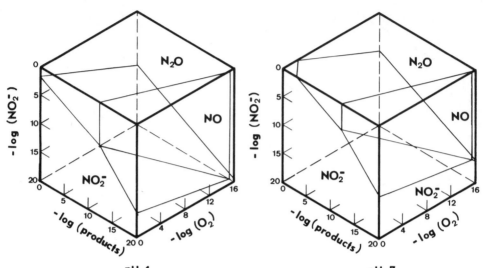

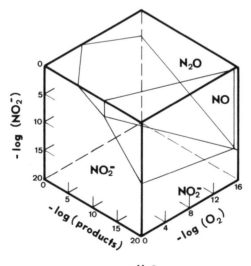

Figure 3.   Representation of the regions of stable reaction
   products taking into account all reactions except the N_2-
   and NO_2-producing ones.

activities is given in Tables I, II and III for pH 4, 7 and 9,
respectively.   In these tables the most stable products
are given at the top of each table, followed by compounds of
lower stability.   In all cases, except at $(O_2) = 0.2$ with
NO_2, N_2 is the most stable compound.   Irrespective of pH,
the relative stability of NO_2 decreases with decreasing (O_2).

Table I

List of N-Compounds with Decreasing Stability at pH 4
as a Function of (O_2)

$(O_2)=0.2$	$(O_2)=10^{-2}$	$(O_2)=10^{-4}$	$(O_2)=10^{-8}$	$(O_2)=10^{-16}$
NO_2^-	NO_2^-	NO_2^-	NO_2^-	NO_2^-
+	+	+	+	+
N_2 NO_2	N_2	N_2	N_2	N_2
↓	↓	↓	↓	↓
N_2O NO	NO_2 N_2O	NO_2 N_2O	NO_2 N_2O	N_2O
↓	↓	↓	↓	↓
N_2O_4	N_2O_4 NO	N_2O_4 NO	N_2O_4 NO	NO
↓	↓	↓	↓	↓
N_2O_5	N_2O_5	N_2O_5	N_2O_3	NO_2
↓	↓	↓	↓	↓
NO_3 N_2O_3	NO_3 N_2O_3	N_2O_3	N_2O_5	N_2O_3
		↓	↓	↓
		NO_3	NO_3	N_2O_4
		↓	↓	↓
				N_3H
				↓
				$H_2N_2O_2$
				↓
				N_2O_5
				↓
				NH_4^+

It is interesting to note that the most stable N-compounds listed here are the products mostly mentioned as nitrite decomposition products. The calculations show that aerobic conditions are more favorable for NO_2, while anaerobic conditions favor N_2O and NO, with the more N_2O the lower the (O_2). It should be born in mind that the results of thermodynamic calculations do not give any answer to questions on the velocity of obtaining the equilibrium. It is also important to note that the used activity values of the different reaction compounds are values at the point of reaction, while usually in soil or solution experiments overall values are used.

From the obtained theoretical results, it can be concluded that the most stable nitrite decomposition products are the same as the N-compounds generally found in experiments with soils and solutions. Besides molecular nitrogen, NO_2, N_2O and NO are the most stable decomposition products. Oxygen tension as well as pH have a minor influence on their stability relative to each other.

Table II
List of N-Compounds with Decreasing Stability
at pH 7 as a Function of (O_2)

$(O_2)=0.2$	$(O_2)=10^{-2}$	$(O_2)=10^{-4}$	$(O_2)=10^{-8}$	$(O_2)=10^{-16}$
NO_2^-	NO_2^-	NO_2^-	NO_2^-	NO_2^-
$+$	$+$	$+$	$+$	$+$
N_2 NO_2	N_2	N_2	N_2	N_2
↓	↓	↓	↓	↓
N_2O NO	NO_2	N_2O NO_2	N_2O NO_2	N_2O
↓	↓	↓	↓	↓
N_2O_4	N_2O NO	NO	NO	NO
↓	↓	↓	↓	↓
N_2O_5	N_2O_4	N_2O_4	N_2O_4	NO_2
↓	↓	↓	↓	↓
N_2O_3 NO_3	N_2O_5	N_2O_5	N_2O_3	N_2O_3
	↓	↓	↓	↓
	N_2O_3 NO_3	N_2O_3	N_2O_5	N_2O_4 N_3H
		↓		
		NO_3		

Table III
List of N-Compounds with Decreasing Stability
at pH 9 as a Function of (O_2)

$(O_2)=0.2$	$(O_2)=10^{-2}$	$(O_2)=10^{-4}$	$(O_2)=10^{-8}$	$(O_2)=10^{-16}$
NO_2^-	NO_2^-	NO_2^-	NO_2^-	NO_2^-
$+$	$+$	$+$	$+$	$+$
N_2 NO_2	N_2	N_2	N_2	N_2
↓	↓	↓	↓	↓
N_2O NO	NO_2	NO_2	N_2O NO_2	N_2O
↓	↓	↓	↓	↓
N_2O_4	N_2O NO	N_2O NO	NO	NO
↓	↓	↓	↓	↓
N_2O_5	N_2O_4	N_2O_4	N_2O_4	NO_2
↓	↓	↓	↓	↓
NO_3	N_2O_5	N_2O_5	N_2O_3	N_2O_3
↓	↓	↓		↓
N_2O_3	N_2O_3 NO_3	N_2O_3		N_2O_4 N_3H
		↓		
		NO_3		

REFERENCES

Bezdicek, D. F., J. M. MacGregor and W. P. Martin. "The Influence of Soil-Fertilizer Geometry on Nitrification and Nitrite Accumulation," *Soil Sci. Soc. Am. Proc.* 35:997-1002 (1971).

Bollag, J. M., S. Drzymala and L. T. Kardos. "Biological versus Chemical Nitrite Decomposition in Soil," *Soil Sci.* 116:44-50 (1973).

Bulla, L. A., C. M. Gilmour and W. B. Bollen. "Nonbiological Reduction of Nitrite in Soil," *Nature* 225:664 (1970).

Chao, T. T. and W. Kroontje. "Inorganic Nitrogen Oxidations in Relation to Associated Changes in Free Energy," *Soil Sci. Soc. Am. Proc.* 27:44-47 (1963).

Chapman, H. D. and G. F. Liebig. "Field and Laboratory Studies of Nitrite Accumulation in Soil," *Soil Sci. Soc. Am. Proc.* 16:276-282 (1952).

Justice, J. K. and R. L. Smith. "Nitrification of Ammonium Sulfate in a Calcareous Soil as Influenced by Combination of Moisture, Temperature and Levels of Added Nitrogen," *Soil Sci. Soc. Am. Proc.* 26:246-250 (1962).

Nelson, D. W. and J. M. Bremner. "Factors Affecting Chemical Transformations of Nitrite in Soils," *Soil Biol. Biochem.* 1:229-239 (1969).

Nelson, D. W. and J. M. Bremner. "Gaseous Products of Nitrite Decomposition in Soils," *Soil Biol. Biochem.* 2:203-215 (1970).

Smith, J. H. "Relations Between Soil Cation Exchange Capacity and the Toxicity of Ammonia to the Nitrification Process," *Soil Sci. Soc. Am. Proc.* 28:640-644 (1964).

Tyler, K. B. and F. E. Broadbent. "Nitrite Transformations in California Soils," *Soil Sci. Soc. Am. Proc.* 24:279-282 (1960).

Van Cleemput, O. and L. Baert. "Gibbs Standard Free Energy Changes of Different Nitrogen Reactions as Corrected for Biological Real pH and O_2 Activities," *Meded. Faculteit Landbouww. Gent.* 39:1-28 (1974).

Van Cleemput, O. and L. Baert. "Theoretical Considerations on Nitrite Self-Decomposition Reactions in Soils," *Soil Sci. Soc. Am. J.* 40:322-323 (1976).

Van Cleemput, O., W. H. Patrick Jr. and R. C. McIlhenny. "Nitrite Decomposition in Flooded Soil under Different pH and Redox Potential Conditions," *Soil Sci. Soc. Am. J.* 40:55-60 (1976).

Wullstein, L. H. and C. M. Gilmour. "Nonenzymatic Gaseous Loss of Nitrite from Clay and Soil Systems," *Soil Sci.* 97:428-430 (1964).

Wullstein, L. H. and C. M. Gilmour. "Nonenzymatic Formation of Nitrogen Gas," *Nature* 210:1150-1151 (1966).

CHAPTER 48

NITROGEN LOSS BY LEACHING FROM AGRICULTURAL FIELDS, RELATED TO MICROENVIRONMENTAL FACTORS

GÖTE BERTILSSON

Bulstofta Experiment Station
Supra AB
Landskrona, Sweden

INTRODUCTION

In agricultural areas the nitrogen from agricultural fields constitutes a major part of the input to aquatic systems. The size of this component of the global N-cycle has been estimated by various methods, primarily based on measurements in river systems (Ahl and Oden, 1972; Omernik, 1976). Studies of microenvironments, in this case agricultural fields or still smaller units, not only contribute to this knowledge but furthermore give information concerning other important questions: how the leaching of nitrogen is influenced by soil and climatic factors as well as agricultural practices. This aids in the interpretation of existing measurements in aquatic systems, in predicting trends caused by changes in land use and agricultural practices, and in establishing guidelines concerning suitable agricultural practices.

Studies on leaching of nitrogen (nitrate) in relation to factors as N-intensity, type of fertilizer, crop, soil and soil profile characteristics have been performed at Bulstofta experiment station in southern Sweden (Supra AB).

METHODOLOGY

1. *Field experiments* with different agricultural treatments combined with drainage measurements. Large field plots (about 20 x 80 m) are arranged according to the existing tile drain system so that one drain represents the plot. Sampling of drain water and flow measurements has been performed at intervals.

Characteristics. Represents a true microenvironment. Rather expensive and area consuming. Gives
few, but reliable, data concerning differences in
nitrogen outflow caused by treatments or climatic
conditions.

2. *Lysimeter experiments*

Characteristics. May be regarded as a model of a
microenvironment with certain characteristics. Can
be performed inexpensively enough to permit extensive
investigations concerning the influence of different
factors. Care must be taken when interpreting the
data to field conditions.

The treatments are presented in the **result** section below.
    In the experiments presented here, no attempt has been
made to use undisturbed soil columns in the lysimeters. Soil
material from appropriate depths has been used. As far as the
topsoil is concerned this represents the normal conditions
after tilling operations. The subsoil, however, has a somewhat higher porosity than normal and is probably a more optimal growth environment.

RESULTS

Field Experiments

    Gluggstorp, Landskrona.
    Started 1972.

Soil:	Glacial loam Excellent agricultural productivity
Treatments:	1. Optimal fertilization as judged by farmer. 2. As 1 + 50 kg N/ha
Drainage measurements:	Sampling for chemical analysis and flow measurement about once a week during periods of drainage
Results:	The parameter best reflecting the difference between treatments is the nitrate content of the drain water. Together with the **cropping** sequence and fertilization this is shown in Figure 1. Two replicates were performed. The nitrogen balance is presented in Table I.

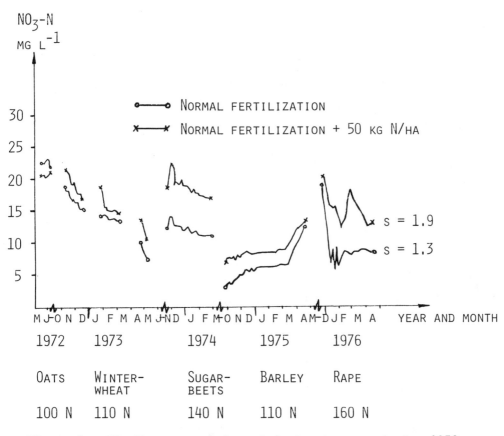

Figure 1. NO_3-N concentration of drainage water during 1972–
1976. Mean of two replicates, standard deviation(s),
crops and "normal fertilization" each year.

Table I
Components of the Nitrogen Balance[a]

	Fertilization			
	Normal		Normal + 50 kg N	
Fertilizer	+115		+165	
Crop Uptake		−128		−157
Leaching		− 22		− 29
	+115	−150	+165	−186
Crop residues returned	+ 44		+ 60	
	+159	−150	+225	−186
Balance (sub−)	+9		+39	

[a]Kg N/ha and year, mean 1972–1975 (compare Figure 1).

Leaching Experiment

Leaching of nitrogen as influenced by nitrogen fertilization, different crops, management practices and soil profile
characteristics are summarized in Figure 2 and Tables II and
III. The nitrogen form is used as a variable to a higher
extent than is shown by this presentation, but in general there

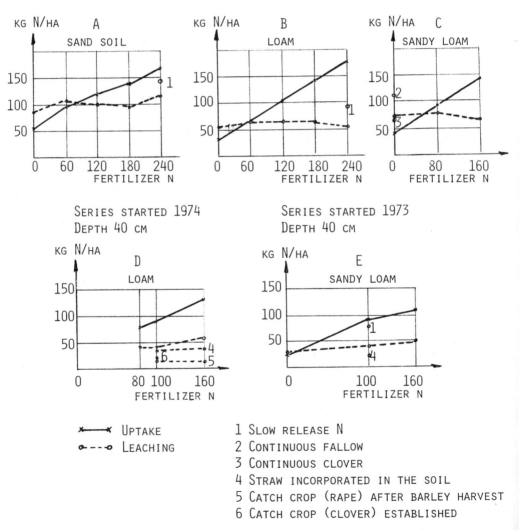

Figure 2. Lysimeter experiments, nitrogen. Crop uptake
(grain + straw) and leaching. Mean of two replicates,
standard deviation(s). Data for 1975-1976 (crop 1975:
barley).

is no difference depending on whether nitrate-, ammonium- or urea-N has been used as fertilizer in these experiments. However, in additional experiments with irrigated sandy soils the leaching losses were larger with nitrate-N than with ammonium- or urea-N.

The straw from cereal crops normally has been removed. The effects of incorporation in the soil after harvest are studied in some of the presented lysimeter series.

Sources of Error

The leachate is collected at intervals, six to ten times a year. The possibility of nitrate reduction and other reactions in the collecting flasks cannot be neglected. However, experience shows first that in experiments with added nitrate to soil without vegetation the recovery in the leachate is good, second that no unexpected differences between sampling times occur, and third that the lysimeters generally give high nitrate leaching as compared to true field systems.

DISCUSSION

There is an enormous amount of information available concerning the reactions of nitrogen in agricultural soils (Viets, 1974; Kolenbrander, 1975). However, the knowledge of rates and quantities is incomplete and sometimes the results of different experiments seem to be contradictory.

The data presented above are in good agreement with the general knowledge (*cf.* Lindhard, 1975; Jung and Jürgens-Gschwind, 1975). When the experiments are classified into different groups according to the characteristics of the microenvironment, the results fit into a scheme and the seeming contradictions disappear (Table IV). The characteristics used in Table III are those that seem to be most important in southern Sweden.

Where do the field experiments fit into this scheme? Although the Gluggstorp experiment presented above is located on a site with excellent productivity, there is an increase in leaching at the high (over optimal) fertilizer level. Thus, it seems not to belong to the first category above, rather to the second. Lysimeters can probably represent an environment too ideal for generalizing to practical agriculture.

A field experiment in Denmark (Kjellerup, 1975) with the N-rates 0, 55, 110 and 165 kg N/ha has given similar results. The leaching is affected only a little when the N-intensity to cereal crops increases from 0 to 55 and even to 110, and more with the next step to 165 (optimum intensity: about 100).

Table II
Lysimeter Experiment with the Variable Soil Profile,
Depth and Crop Type[a]

Nitrogen Balance Components;
Mean of Two Replicates Expressed as kg N/ha and Year

Lysimeter Depth (m)	Crop	Crop Uptake	Leaching
1.2	Oats	165	8
0.8	Oats	163	22
0.4	Oats	141	39
1.2	Lettuce	158	30
0.8	Lettuce	156	46
0.4	Lettuce	143	50
Standard deviation		6.3	7.1

[a]Soil: glacial loam. Nitrogen fertilizer: Calcium nitrate corresponding to 200 kg N/ha. Experiment started 1974. Same crops 1975.

Table III
Lysimeter Experiment Started 1976[a]

Treatment Fertilizer N (kg/ha)	Crop	Uptake (kg N/ha)	Content of NO_3-N in drain water (mg/l)
0	Barley	39	19
60	Barley	78	17
120	Barley	108	29
120	Barley, straw plowed in	(108)	7
120	Barley, rape as cover crop	108	<0.5
60	Barley, clover as cover crop	79	8
120	Sugar beet	115	3
0	Beans (Vicia)	214	59
Standard deviation		3.7	2.4

[a]Depth 40 cm. (Topsoil lysimeter). Main crop: barley; soil: loam; variables: fertilizer type, rate and time of application, crop and cultural practice, only partly presented. The leaching in kg/ha cannot yet be calculated. Presented is the average nitrate-N content in the first 40 mm drained.

Table IV
Lysimeter Experiments Grouped According to the
Characteristics of the Microenvironment

Examples	Characteristics	Function
Figure 2, b and c	Deep profile Crops with deep root systems Ideal crop stand and soil structure Moderate humidity	No relation between nitrogen intensity in the system and leaching in the range studied (twice normal intensity)
Figure 2, a Table II	As above, but less optimal root development, caused by soil factors (sand) or crop factors (lettuce)	Increased leaching at high nitrogen intensity
Figure 2, d and e Table III	Shallow profiles Moderate humidity	As above
Data not presented here	Shallow profiles High humidity (irrigation)	Possible direct relation between N-intensity and leaching In contrast to other systems: ammonium-N reduces leaching as compared to nitrate Possible advantage for slow release N

Some other treatments as represented in Figure 2 and Tables II and III are:

- ploughing in of straw after harvest has markedly reduced the leaching of nitrate, at least at high nitrogen intensities

- a catch crop established after harvest has, during favorable conditions, changed the N-economy of the environment completely. The leaching of nitrate has been much reduced. Rape has been more efficient than clover, but also during the phase of active growth, clover has been better than no vegetation.

- continuous clover has given about the same nitrate leaching as fertilized cereal crops (higher content but less amount of water). After beans the nitrate concentrations in the drain water have been very high.

- sugar beets, even well fertilized, have given small outflow of nitrate. Probable cause: the vegetative phase lasts late in the autumn.

- the slow release nitrogen sources tested (glycolurea, sulfur-coated urea) have in these experiments caused increased nitrate leaching compared to conventional fertilizers. Probable cause: they have delivered nitrogen also during periods without vegetation.

REFERENCES

Ahl, T. and S. Oden. "River Discharge of Total Nitrogen, Total Phosphorus and Organic Matter into the Baltic Sea from Sweden," *Ambio spec. Rep.* 1:51-56 (1972).

Jung, J. and S. Jürgens-Gschwind. "Die Stickstoffbilanz des Bodens, dargestellt an Lysimeterversuchen," *Landw. Forsch. Sonderheft 30/II*, 57-77 (1974).

Kjellerup, V. "Kvaelstofgodskningens inflytande pa dranvandets inhold av nitratkvaelstof (The Influence of Nitrogen Fertilization on the Content of Nitrate Nitrogen in Drain Water)," *Statens Forsogsvirksomhed i Plantekultur, 1220* Meddelelse, Askov, 6600 Vehen, Danmark (in Danish).

Kolenbrander, G. J. "Nitrogen in Organic Matter and Fertilizer as a Source of Pollution," *Conference on Nitrogen as a Water Pollutant, Proc. IAWPR*, Vol. 1. (Copenhagen, 1975).

Lindhard, J. "Nitrogen in Crops and Drainage Water.... Lysimeter Experiment, 1962-72," *Tidskrift Planteavl* 74: 536-544, (1975) (Danish, English summary).

Omernik, James M. "The Influence of Land Use on Stream Nutrient Levels," *Ecological Research Series*, EPA-600/3-76-014 (USA) (1976).

Viets, F. G., Jr. "Animal Wastes and Fertilizers as Potential Sources of Nitrate Pollution of Water," in *Effects of Agricultural Production on Nitrates in Food and Water with Particular Reference to Isotope Studies, Proc.* (Vienna:IAEA, 1974), pp. 63-76.

SECTION IV

INTERFACES AND SORPTION

CHAPTER 49

THE ROLE OF INTERFACES IN SOIL MICROENVIRONMENTS

K. C. MARSHALL
N. A. MARSHMAN

School of Microbiology
The University of New South Wales
Kensington 2033, N.S.W.
Australia

INTRODUCTION

A wide range of biogeochemical activities are catalyzed by microorganisms in soils. The rate of any particular reaction varies from soil to soil because of the different physical, chemical and biological characteristics of the soils. In the past, little consideration has been given to the nature of the actual microenvironments that determine the nature and extent of microbially catalyzed reactions within different soils. Here we intend to discuss the role of interfaces in determining the microenvironment of naturally occurring soil microorganisms.

SOIL MICROENVIRONMENTS

A soil microenvironment may be defined as a restricted, but variable, volume of the soil wherein specific or unique reactions affect or are affected by a particular process. We wish to emphasize the microenvironment of individual microorganisms or of microcolonies of microorganisms. In this sense, a microenvironment cannot be defined in terms of a definite fixed volume representing the zone of influence of all factors on the individual or of the individual on its surroundings.

Consider a simplified model (Figure 1) of the many sites of a soil ped. Most microorganisms only function within the aqueous phase situated between solid surfaces and voids in freely draining soils. Filamentous microorganisms are

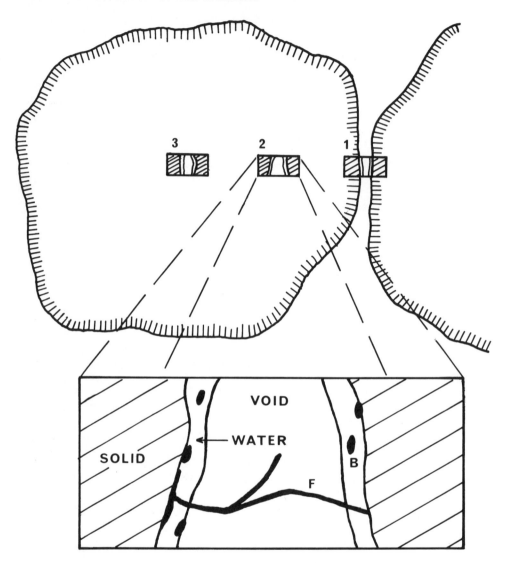

Figure 1.   A schematic representation of soil peds, showing
   the location of microorganism relative to voids and to
   solid surfaces on the ped surfaces (site 1) and within a
   ped (sites 2 and 3).   B and F represent bacteria and a
   filamentous microorganism, respectively.

different in that they are able to bridge across voids.
Regardless of the precise location of a microorganism within
the aqueous phase, the region of influence of a nutrient, a
growth factor of a gas upon the microorganism will be different
from the region of influence of other nutrients, growth
factors or gases.   Also, the region of influence of a specific environmental factor is dependent upon the precise

location of the microorganism in the soil (*e.g.*, Figure 1 -
sites 1, 2 and 3). Such variations in the size of zones of
influence result from differences in solubility, diffusion
and/or flow rate, site of origin, or activity of adjacent
microorganisms. In addition, the zone of influence will
depend upon the biochemical properties of the microorganisms
in question.

Other factors being equal, the nature and activity of
microorganisms in the three sites indicated (Figure 1) within
the soil ped will be influenced by oxygen availability
(Greenwood, 1962; Hattori, 1973; Hattori and Hattori, 1976).
Oxygen should not be limiting in the void at site 1, but in
moist peds oxygen should become progressively limiting on
moving from sites 1 to 3. Aerobic processes should be
favored at site 1, whereas anaerobic processes generally should
be favored at site 3 (Greenwood, 1962). An alternating aerobic-anaerobic cycle might be predicted at site 2 on the basis
of the oxygen-ethylene system postulated by Smith and Cook
(1974) and Smith (1976). Ethylene production by anaerobic
microorganisms at site 3 will diffuse outwards and arrest
any aerobic activity at site 2. As aerobic activity diminishes,
oxygen demand at this site decreases and oxygen will diffuse
further into the internal voids. This oxygen will slow
the anaerobic activity and arrest ethylene production, there-
by restoring the aerobic sequence at site 2, and the cycle
is then repeated.

INTERFACES AND SOIL MICROENVIRONMENTS

An interface is the boundary between two phases in any
heterogenous system. Soils are heterogenous systems consisting
of a variety of solid-liquid, gas-liquid and solid-gas interfaces. Solid-liquid and gas-liquid interfaces are of major
significance in determining microbial activity in soil microenvironments. Solid-gas interfaces occur in relatively dry
soil sites where microbial activity is limited mainly to
filamentous microorganisms capable of growth at low matric
water potentials (Griffin, 1972). Survival of desiccated
microorganisms can be influenced by the nature of potentially
protective colloids or macromolecules sorbed to the solid
surface, the relative humidity of the gas phase, and the nature
of the gas phase (Marshall, 1976). The significance of
interfaces in microbial ecology has been discussed by
Marshall (1975, 1976) and by Hattori and Hattori (1976).

Gas-liquid interfaces are often ignored, but they do
represent important features of soil microenvironments
(Figure 2). Motile microorganisms can exhibit either an
aerotactic or a chemotactic response to gas-liquid interfaces,
and bacteria with hydrophobic surfaces accumulate at the
interface as a result of being rejected from the aqueous

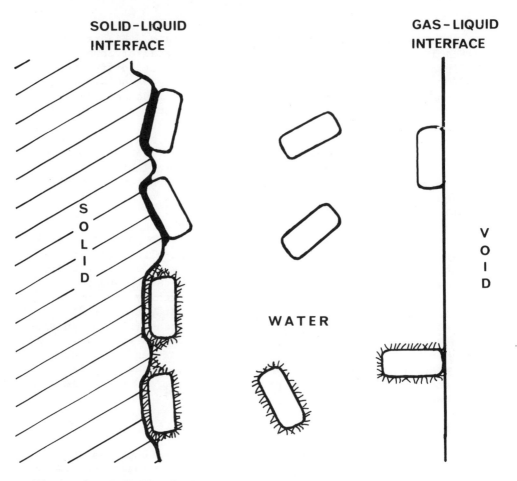

Figure 2.   Detail of sites occupied by microorganisms within
  the aqueous phase of soil.   In addition to those microorganisms located within the bulk aqueous phase, microorganisms accumulate at the gas-liquid interface or firmly
  adhere at the solid-liquid interface.   Colloidal materials
  (clay, organic matter), where present in suspension, are
  indicated as being attracted to the surfaces of microorganisms.

phase (Marshall and Cruickshank, 1973).   Studies in aqueous
systems have demonstrated the preferential accumulation of
organic and inorganic nutrients in "surface microlayers" at
the gas-liquid interface (Parker and Barsom, 1970), resulting
in enhanced microbial activity compared to that in the bulk
aqueous phase (Hatcher and Parker, 1974).

    Solid-liquid interfaces play a major role in defining
the microenvironment of individual microorganisms in soils.
When the nutrient level in the aqueous phase is limited, the
solid-liquid interface provides a site for nutrient concentration and increased microbial activity.   Microorganisms are

attracted to surfaces by a variety of physicochemical mechanisms (Marshall *et al.*, 1971; Marshall 1975, 1976; Hattori and Hattori, 1976), and may be reversibly or irreversibly sorbed to the solid surface. Irreversible sorption involves an anchoring of the cells to the surface by a process of polymer bridging (Figure 2). The nature of a solid surface will control the extent to which nutrients and microorganisms are attracted to the surface.

Soils represent a solid matrix within which voids, water and microorganisms are found. Soils are almost infinitely variable with respect to the nature and admixture of the solid components. The complexity of the solid components surrounding even a single void system is illustrated in Figure 3A. These components represent potential microbial

Figure 3. A: SEM illustrating the complex organization of soil particulate materials surrounding a soil void.
B: Detail of the area indicated in A, showing limited microbial colonization of the surfaces.

microenvironments of varying quality. The limited extent of colonization of one of the surfaces is shown in Figure 3B. Even this surface is far from homogenous and is generally indicative of the solid interfaces encountered in soil microenvironments. The three-dimensional organization of particulate components of soils is referred to as the soil fabric (Brewer, 1964), and definition of soil microenvironments in the language of the soil micromorphologist is essential for a thorough understanding of microbial behavior in soils.

A very simple example of modification to microbial activity as a result of soil fabric differences can be illustrated by reference to Figure 4. Sand grains from a sandy soil

Figure 4.   SEMs of sand grains: *A*, from a sandy soil; *B*, from a podzolic soil; *C*, detail of A, showing the smooth quartz surface; *D*, detail of B, showing kaolinitic clay on the sand grain surface.

and a podzolic soil were compared under the scanning electron microscope and the individual sand grains, although of comparable size and shape (Figures 4A and B), represent entirely different surfaces at the microbial level (Figures 4C and D). One of the sands (Figure 4C) has a smooth quartz surface, whereas the other has a kaolinitic clay firmly fixed to it (B. Davey, University of Sydney, personal communication). In the presence of added nutrients, the sand with the smooth quartz surface was toxic to a number of bacteria, but in the clay-coated sand vigorous growth of the bacteria was observed.   This example emphasizes the role of surface features in modifying microbial behavior.   In a heterogenous soil situation, such as that shown in Figure 3A, an accurate description of the micromorphological features of all surfaces must be necessary for accurately defining the various micro-environments available to microorganisms.

Further complications are introduced where colloidal material is suspended in the aqueous phase, even in minute quantities. This colloidal material adsorbs to microbial surfaces (Marshall, 1976) and forms an envelope around the cells (Figure 2). This envelope may protect the bacteria from desiccation (Marshall, 1964) and predator-prey inter-actions (Roper and Marshall, 1974), but it may also alter the metabolic activity and population balance of soil microorgan-isms. General aspects of the affects of clay minerals on the metabolism of soil microorganisms have been reviewed by Stotzky (1972) and Filip (1973).

ACKNOWLEDGMENTS

Original investigations reported in this paper have been supported by grants from the Australian Research Grants Committee.

REFERENCES

Brewer, R. *Fabric and Mineral Analysis of Soils* (New York: J. Wiley & Sons, 1964).

Filip, Z. "Clay Minerals as Factors Influencing Biochemical Activity of Soil Microorganisms," *Folia Microbiol.* 18: 56–74 (1973).

Greenwood, D. J. "Nitrification and Nitrate Dissimilation in Soil. II. Effect of Oxygen Concentration," *Plant Soil* 17:378–389 (1962).

Griffin, D. M. *Ecology of Soil Fungi* (London: Chapman and Hall, 1972).

Hatcher, R. F. and B. C. Parker. "Microbiological and Chemical Enrichment of Freshwater-Surface Microlayers Relative to the Bulk-Subsurface Water," *Can. J. Microbiol.* 20:1051–1057 (1974).

Hattori, T. *Microbial Life in the Soil* (New York: Marcel Dekker, Inc., 1973).

Hattori, T. and R. Hattori. "The Physical Environment in Soil Microbiology: An Attempt to Extend Principles of Microbiology to Soil Microorganisms," *CRC Critical Rev. Microbiol.* 4:423–461 (1976).

Marshall, K. C.   "Survival of Root-Nodule Bacteria in Dry Soils Exposed to High Temperatures," *Aust. J. Agric. Res.* 15:273-281 (1964).

Marshall, K. C.   "Clay Mineralogy in Relation to Survival of Soil Bacteria," *Ann. Rev. Phytopathol.* 13:357-373 (1975).

Marshall, K. C.   *Interfaces in Microbial Ecology* (Cambridge, Massachusetts: Harvard University Press, 1976).

Marshall, K. C. and R. H. Cruickshank.   "Cell Surface Hydrophobicity and the Orientation of Certain Bacteria at Interfaces," *Arch. Mikrobiol.* 91:29-40 (1973).

Marshall, K. C., R. Stout and R. Mitchell.   "Mechanism of the Initial Events in the Sorption of Marine Bacteria to Surfaces," *J. Gen. Microbiol.* 68:337-348 (1971).

Parker, B. and G. Barsom.   "Biological and Chemical Significance of Surface Microlayers in Aquatic Ecosystems," *Bioscience* 20:87-93 (1970).

Roper, M. M. and K. C. Marshall.   "Modification of the Interaction Between *Escherichia coli* and Bacteriophage in Saline Sediment," *Microbial Ecology* 1:1-13 (1974).

Smith, A. M.   "Ethylene in Soil Biology," *Ann. Rev. Phytopathol.* 14:53-73 (1976).

Smith, A. M. and R. J. Cook.   "Implications of Ethylene Production by Bacteria for Biological Balance of Soil," *Nature* (London), 252:703-705 (1974).

Stotzky, G.   "Activity, Ecology and Population Dynamics of Microorganisms in Soil," *CRC Critical Rev. Microbiol.* 2:59-137 (1972).

CHAPTER 50

POLYCHLORINATED BIPHENYLS (^{14}C) IN SOILS: ADSORPTION, INFILTRATION, TRANSLOCATION AND DECOMPOSITION

H. W. SCHARPENSEEL*
B. K. G. THENG**
S. STEPHAN***

Ordinariat für Bodenkunde der Universität Hamburg
Institut für Bodenkunde der Universität Bonn
Bonn, Federal Republic of Germany

INTRODUCTION

Because of their chemical and thermal stability, polychlorinated biphenyls (PCBs) have been widely used in a variety of industrial applications, notably the electrical industry where they function as a heat-transfer and dielectric fluid (*e.g.*, Edwards, 1971). Therefore these compounds and their residues have found their way into the marine and terrestrial environment and, being fat-soluble, have entered the food chain and accumulated in the adipose tissue of fish, birds and mammals including humans (Jensen *et al.*, 1969; Koeman *et al.*, 1969; Hammond, 1972; Risebrough *et al.*, 1972; Yobs, 1972; Solly and Shanks, 1974; Harvey and Steinhauer, 1975). Indeed, PCBs, together with the organochlorine pesticides (chlordane, DDT, dieldrin, lindane) have become the most widespread and abundant synthetic pollutants of the global ecosystem (Risebrough *et al.*, 1968). The occurrence and persistence of PCBs in and their impact on the environment have been discussed by a number of workers (*e.g.*, Gustafson, 1970; Peakall and Lincer, 1970; Nelson *et al.*,

---

*Present address: Ordinariat für Bodenkunde der Universität Hamburg, 2057 Reinbek, Schloss, FRG.
**Present address: Soil Bureau, Department of Scientific and Industrial Research, Private Bag, Lower Hutt, New Zealand.
***Present address: Institut für Bodenkunde der Universität Bonn, 53 Bonn, Nußallee 13.

1972) and have formed the theme for a national conference in
1975, sponsored by the United States Environmental Protection
Agency. Incidentally, the toxic and possible teratogenic
effects of PCBs on humans in Japan and the United States
have recently made news headlines.

Despite the considerable amount of attention and effort
that has been devoted to the various aspects of environmental
contamination by PCBs, surprisingly little is known about the
fate of these compounds in the soil system. The available
evidence indicates that PCBs, more so than the organochlorine
pesticides, are effectively immobilized when they are added
to soil (Weil *et al.*, 1972; 1973; Haque *et al.*, 1974). The
adsorption and retention of PCBs by soil and soil constituents
are influenced by the number of chlorine atoms in the molecule,
the more highly chlorinated derivatives being the more tena-
ciously held (Tucker *et al.*, 1975; Haque and Schmedding, 1976).
The decomposition of PCBs by microorganisms has been studied
by Kaiser and Wong (1974) as well as by Baxter *et al.*, (1975),
who observed a similar trend in that the rate of degradation
decreased with an increase in the chlorine content of the
molecule.

The present study was carried out to gain further information on (A) the interaction of PCBs with specific soil
constituents, (B) the infiltration and translocation of PCBs
in soil and their retention by soil against percolating water,
(C) the distribution of soil-applied PCBs among various components of the soil ecosystem or biotope, and (D) the decomposition of PCBs in soil under biotic and sterile conditions.
The use of uniformly labelled ^{14}C-labelled PCBs provided a
convenient means of obtaining relatively accurate experimental
data on these aspects of their behavior.

EXPERIMENTAL

(A) The montmorillonite used was the < 2 µm equivalent
spherical diameter fraction of a commercial bentonite sample,
supplied by E. Merck, Darmstadt. Details on the preparation
and characterization of the various cationic forms of the
clay have been given elsewhere (Theng and Scharpenseel, 1975).
The humic (and fulvic) acid samples were obtained from a
sample of a chernozem horizon (Aselerwald) using the classical alkali extraction method (Scharpenseel *et al.*, 1968).
The PCB sample was a ^{14}C-labelled 2,3,4-trichloro derivative.
Adsorption isotherms were determined by equilibrating 100-mg
quantities of the "dry" clay with 10-ml portions of a *n*-hexane
solution of PCB of varying strength for 16 hr at 25° C.
After centrifugation, the activity of the clear supernatant
liquid was measured using a Packard liquid scintillation
spectrometer (Scharpenseel *et al.*, 1977a). The apparent
amount adsorbed was estimated from the difference in count

rate between the initial solution and that at equilibrium with the adsorbent.

(B)   Briefly, the method consisted of percolating water through two undisturbed soil cores contained in glass columns (50 cm long and 10 cm across), to which 50 µC of ^{14}C-PCB (30.3% di- and 69.7% trichloro biphenyl) dissolved in *n*-hexane had been applied, collecting the percolate, and measuring its activity as before.   Water was periodically introduced to the top of each column at a rate equivalent to 100 mm of rainfall per week until a total of 900 mm (column 1) and 1700 mm (column 2) had been applied (Kerpen and Scharpenseel, 1967).   At the end of the leaching experiment, the cores were taken out of the columns and placed in a gypsum case lined with X-ray film to obtain an autoradiograph (14 days exposure). Each core was then dissected at definite intervals along its length and each section was suitably treated for micromorphological analysis using both contact and stripping film autoradiography (Stephan, 1969; Fischer and Werner, 1971).

(C)   Two biotopes of 12 and 24 m^2 areal dimensions were chosen, over which ^{14}C-PCB was uniformly sprayed at a rate of 2 µC m^{-2}.   The distribution of PCB in soil and humus fractions from different horizons within the soil profile as well as in the earthworm population and in parts of the overlying vegetation was monitored over a period of eight to nine months.   Because the permitted amount of PCB applied (in terms of activity) was limited, counting techniques as used in radiocarbon dating of soils (Scharpenseel and Pietig, 1969) had to be modified in order to preserve accuracy.

(D)   The rate of PCB decomposition was measured in terms of the amount of ^{14}CO$_2$ evolved over a period of time (Scharpenseel and Beckman, 1964) using soil samples before and after sterilization, for different water regimes, and under the influence of ultraviolet radiation.

A full description of the methods outlined in (B), (C) and (D) has been given previously (Scharpenseel *et al.*, 1977 a, b).

RESULTS AND DISCUSSION

Figure 1 shows that in the range of 0-0.4 µg PCB/ml solution all the isotherms belong to the C-class, the amount adsorbed increasing linearly with concentration.   This accords with the recent findings by Bykov *et al.*, (1976) that hydrocarbons give rise to linear isotherms with both kaolinite and montmorillonite.   Following Giles *et al.*, (1960), the curves for montmorillonite may be described in terms of a constant partition of solute (PCB) between the bulk and the interlayer solvent (*n*-hexane), that is, of an equilibrium between the concentration of PCB in the bulk (b) and that in the interlayer (i) phase:

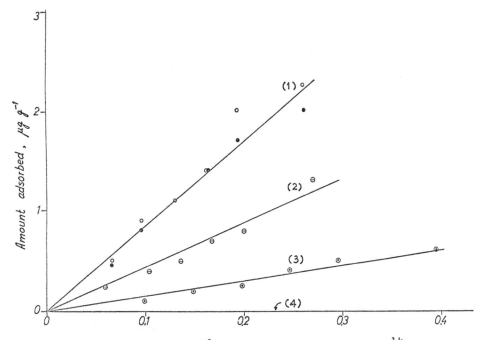

Figure 1.   Isotherms at 25° C for the adsorption of ^{14}C-2, 3,4,-trichloro biphenyl from *n*-hexane by montmorillonite and humic substances.   (1) Ca^{2+}-montmorillonite, 0-0; Cu^{2+}-montmorillonite, 0-0; (2) Na^{+}-montmorillonite; (3) humic acid, air-dry; (4) humic and fulvic acids, dried in the oven at 100° C for 16 hr.

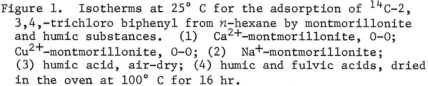

The partition coefficient of the process, K_p, is defined as

$$K_p = (PCB)_i / (PCB)_b$$

$$= S_i/S_b$$

where S_b and S_i refer to the solubility of PCB in the respective phases.

For a given value of $(PCB)_b$ the extent of adsorption for Ca$^{2}\pm$ and Cu$^{2}\pm$ montmorillonite was about twice that for the sodium-saturated clay (Figure 1), indicating that a proportionately greater volume of *n*-hexane was present in the interlayer space of the divalent cation systems. This observation presumably relates to the fact that under ambient conditions of temperature and humidity, the average distance separating individual silicate layers in crystals of Ca$^{2}\pm$ (and Cu$^{2}\pm$) montmorillonite was initially greater as compared with the sodium system. By taking a value of 4.6 A for this distance in the Ca$^{2}\pm$ montmorillonite complex with *n*-hexane

(Eltantawy and Arnold, 1972), the volume (V) of interlayer solvent and hence K_p may be calculated from surface area and adsorption measurements (Scharpenseel *et al.*, 1977a). V_{Ca} = 0.17 ml g^{-1} giving 113 mg g^{-1} as the weight of interlayer n-hexane, which compares well with the value of 110 mg g^{-1} reported by Eltantawy and Arnold.. Similarly, for K_p = 48.7, V_{Na} = 0.09 ml g^{-1} corresponding to a basal spacing of 12.28 Å, which is close to Eltantawy and Arnold's value of 12.37 Å for their Na^+-montmorillonite complex.

The good agreement between the calculated values for different parameters and those obtained previously by independent means lends further support to the suggested mechanism of adsorption. The isotherm for humic acid (HA) may be treated likewise by substituting interstitial for interlayer space as Theng (1972) has proposed for the interaction of alkylammonium ions with soil allophane. That relatively little was adsorbed (V_{HA} = 0.03 ml g^{-1}) would indicate that the interstitial space accessible to n-hexane was limited. By the same token, no interstitial penetration by solvent apparently occurred for humic (and fulvic) acid, which had previously been dried at 100° C. Steric factors might be involved although kinetic effects were probably more important here. Eltantawy and Arnold observed, for example, that the rate of n-hexane intercalation by montmorillonite was strongly influenced by the water content of the system, being markedly reduced for samples that were preheated to 220° C.

It is interesting to note that in an aqueous medium humic acid is a better adsorbent for PCB than clays or whole soils, although the maximum amount adsorbed was comparable to, if not less than, that indicated in Figure 1 because of the very low solubility of PCB in water (Haque and Schmedding, 1976). That the order of adsorbability in n-hexane was the reverse of that obtained in water may partly be explained in terms of competition between solute and solvent for the surface. In an aqueous medium PCB would be preferentially taken up by the hydrophobic surface of humic acid, whereas in a hydrocarbon medium, the solvent would be preferred to the solute. This observation is relevant to the behavior of PCB in soil columns discussed below. In the initial stages PCB, applied as a solution in n-hexane, would be expected to be associated more with the mineral than with the organic fraction of the soil. When most of the solvent has evaporated and water is percolated through, there may well be a local redistribution and translocation of PCB within the system so that it tends to accumulate around humified organic residues (Figure 2).

Table I indicates a perfusion of 0.22 µc, equivalent to 0.44 percent of the total application after nine times 100 mm rainfall simulation with slowly decreasing individual rates. Leaching after 1700 mm of rainfall amounted to

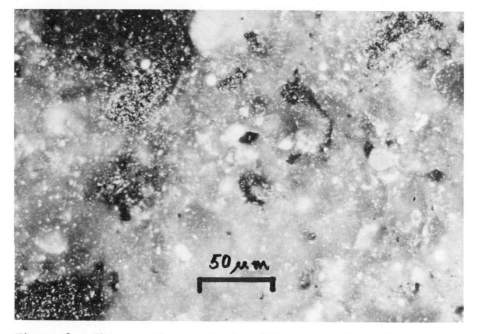

Figure 2.   Thin section stripping film autoradiography
   indicating silver grains concentrated around organic
   rotting products.

0.36 µc, whereby the individual rates diminished rather
constantly. These amounts apply to PCB-application in hexane
solution at 2 cm below the soil surface.

   After excessive moisture drained out, the soil columns
were pushed out of the glass casing and transfered in a gypsum
cast, laid out with X-ray film. Following two weeks of exposure, a ^{14}C-PCB concentration closely adjacent to the
mixing zone, 2 to 10 cm below the upper surface, was revealed.

   The distribution pattern of the labelled PCB in the
soil plasma was investigated by autoradiographic methods.
Thin sections, produced out of the aforementioned soil columns
by freeze during and subsequent soaking under vacuum with
polyester resin, were covered with stripping emulsion and
exposed for sufficient time to obtain autoradiographic marks
of the labelled PCB distribution. Vestopal as well as Araldit
resin have been applied. The Araldit hardens quite fast,
whereas Vestopal requires several weeks of slow polymerization
and induration to give optimal results. Since in no case
traces of activity were recognized in the autoradiographs'
resin-filled holes and pores, translocation of the PCB by
contacts with the resin filling does not occur or is negligible.

   The thin sections produced were protected by dipping
them three times into a 0.5 percent solution of Pioloform in
chloroform. Kodak AR 10 stripping film was used for the autoradiographs as described by Fischer and Werner (1971). Although
a certain amount of scattering of the PCB in the soil has to

Table I
Eluted Amounts of ^{14}C-PCB (μc and percent)

100 mm Simulated Rainfall	Column 1 μc x 10^{-2} (quench-corrected Int.St.)	% of Total Administered Amount	Column 2 μc x 10^{-2} (quench-corrected Int.St.)	% of Total Administered Amount
1st	2.20	0.0439	2.40	0.0480
2nd	2.20	0.0441	2.99	0.0598
3rd	3.09	0.0617	3.33	0.0667
4th	3.52	0.0705	4.99	0.0998
5th	3.09	0.0619	4.23	0.0846
6th	2.21	0.0442	3.25	0.0650
7th	2.44	0.0488	2.93	0.0587
8th	1.75	0.0351	2.42	0.0484
9th	1.58	0.0316	2.54	0.0509
10th			1.56	0.0312
11th			1.12	0.0224
12th			0.74	0.0147
13th			1.06	0.0213
14th			0.65	0.0131
15th			0.79	0.0158
16th			0.54	0.0108
17th			0.66	0.0133
	0.22 μc	0.442 %	0.361 μc	0.725 %

be taken in account, the remaining connection betwen stripping
film and thin section allows clear distinction of the concentration points of PCB activity (dark silver grains) with the
aid of the microscope. Based on these investigations there
is but little activity throughout the soil plasma, consisting
of the inorganic, organic and clay organic soil colloids,
which are ordinarily considered to be the main sink of adsorbed
substances in the soil. Clay aggregates (relictic argillans
of illite) show little activity. Instead of this, clouds of
silver grains have been found covering more or less altered
plant remnants, rotting products of the preceding field
crops. (Figure 2).

High activity by PCB attachment is revealed in weakly
browned tissues showing well-formed cells, in compressed as
well as in more darkened plant tissues, in coal-like remnants
and in dark plant matter altered by metabolic breakdown processes. The attraction forces of these rot products to a
lipophilic substance such as PCB is underlined by the fact
that the monostyrene solvent accompanying the Vestopal application, and in which the PCB is soluble, was not capable of
incorporating detectable amounts of PCB out of these rotting
products in the course of the several weeks of resin hardening.

Although the adsorption isotherms of PCB in Chapter 2
show clearly that PCBs can be retained and fixed by clay
minerals, from the results of the autoradiographs one could
conclude that the greatest part of the PCB is retained by only
a small fraction of organic matter in various stages of decomposition. In consequence, the contamination of the greatest part of the soil would be minimal. The relatively high
concentration of the PCB in the rotting products, however,
could enhance the transfer of the PCB into the biosphere
because the weakly altered plant remnants are readily taken
up by the soil fauna and represent an important part of the
food chain.

For the PCB inventory (infiltration and translocation)
two small biotopes were used: first, 24 m^2 of Hapludalf
(fenced experimental station of Kleinaltendorf), second 12 m^2
sandy Haplumbrept (garden of Institute at Reinbek).

1) Kleinaltendorf biotope: each of the four lots,
2 x 3 m, received 12.5 µc of ^{14}C-PCB, emulsified with Tween
80 in water and sprayed uniformly over the area (Figure 3).

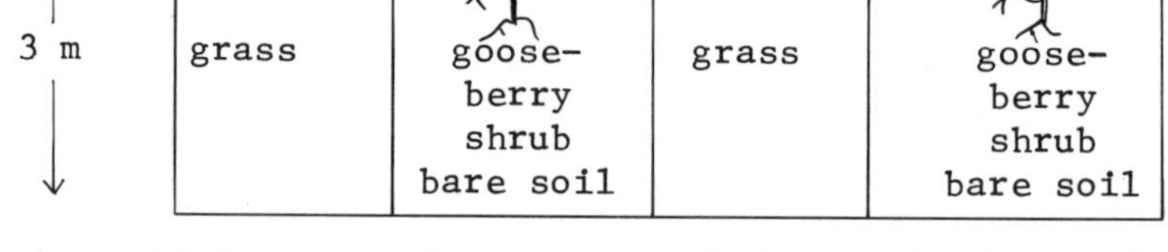

Figure 3.   Subdivision of experimental lot Kleinaltendorf.

With a sprinkler, 4 mm of water were added initially. The Hapludalf soil had clay infiltration, no free carbonates and < 2 percent organic matter. Each .sampling process included two treatments (two grass lots and two lots besides a gooseberry-shrub free of vegetation due to herbicide treatment). In addition a subdivision in three horizons (Ap 0-30, Al 30-60, Bt 60-90 cm) and a fractionation into whole soil, 2-0.06 mm, < 0.06 mm fractions, humic acids + humines, fulvic acids, humin coal (the remaining C after humine extraction), hexane extract, residue after hexane extraction were included. Also rainworms were collected, combusted and tested for ^{14}C-PCB, just as the other fractions. Each sampling process thus comprised two plant samples (1 x grass + 1 x gooseberry leaves and small branches) and 48 soil fractions, all sample-C subject of conversion into benzene, with following low-level measurement.

Total rainfall during the observation period was 504.6 mm. Table II reflects the natural ^{14}C concentrations before administration of the ^{14}C-PCB, *i.e.*, the blank. The alkali-treated humic matter fractions have attracted bomb carbon from the atmosphere. The other fractions are expectedly of the order of 6 to 9 cpm, compared with 10.5 cpm for the National Bureau of Standard's NBS-95 percent contemporary oxalic acid C-standard.

The excess of ^{14}C-PCB activity beyond the blank level, revealed in Table III is due to PCB and PCB metabolite incorporation in the various fractions.

Results from Tables II and III:

- C-contents remained constant in course of samplings of June 25 and August 6. They did not need further repetitions.
- Hexane Soxhlet extractions of PCB from air dry soil are not quantitative.
- PCB contents are higher in the texture fraction < 0.06 mm compared with fraction > 0.06 mm.
- Humic matter fractions show relatively high PCB sorption, but based on 1 g C not more than the total soil, which suggests considerable bonding of PCB with nonhumic organic constituents.
- More ^{14}C-PCB is attached to humic acid + humine C than to fulvic acid C. Almost the same level is observed with the humus coal fractions, confirming the strong PCB fixation, which resisted even hot alkali and acid treatment.
- After three weeks PCB distribution reaches to the bottom of the argillic Bt horizon, with a decrease in the lessive Al horizon in the case of the grass-covered lots. After two months little has changed in the bare soil lots, except for reduced ^{14}C-PCB concentration in the Al-horizon. Under grass cover only

Table II
Natural ^{14}C Activities in the Soil
of Experimental Station Kleinaltendorf
before PCB Application (6/5/74)

Sample Material	% C	cpm/g C
Ap 0–25 cm, Total Soil	1.0	9
2–0.06 mm	1.2	8
< 0.06 mm	0.9	10
total humic matter	8.0	25
fulvic acid	4.5	23
humic coal	0.4	17
hexane extract		4
extracted soil	1.1	8
Al 25–50 cm, Total Soil	0.6	8
2–0.06 mm	0.6	7
< 0.06 mm	0.5	9
total humic matter	33.0	10
fulvic acid	4.0	12
humin coal	0.2	13
hexane extract		4
extracted soil	0.6	8
Bt 1.50–75 cm, Total Soil	0.4	7
2–0.06 mm	0.5	7
< 0.06 mm	0.4	8
total humic matter	23.8	6
fulvic acid	3.7	13
humin coal	0.1	7
hexane extract		6
extracted soil	0.4	7
Bt 2.75–110 cm, Total Soil	0.4	8
2–0.06 mm	0.3	7
< 0.06 mm	0.3	7
total humic matter	20.8	8
fulvic acid	4.2	25
humin coal	0.1	11
hexane extract		8
extracted soil	0.3	6

the ochric epipedon is still on level; Al and Bt
horizon approach the blank. This tendency is enhanced
in all lots after four months. Eight months after the
PCB application the specific activity in the Ap
epipedons is also reduced to a fraction, and the subsoil horizons are practically on blank level. The
specific activity in the grass samples (grass was cut
and removed after each sampling) and gooseberry samples
parallels this trend; rainworms from sampling October
9 showed about the same level.

- Based on an annual 5000 kg dry matter yield/ha, *i.e.*,
200 g c/m^2, the experimental area with 24 m^2 would
yield 4800 g C. Assuming an average of 150 cpm/g C,
a total of 720,000 cpm would be removed from the
biotop by the vegetation. Adjusted for 75 percent
counting efficiency, this equals 10^6 dpm, *ca.* 1/2
μc, 1 percent of the administered ^{14}C-PCB. The PCB
remaining after nine months in the Ap epipedon comes
close to 6 μc = 12 percent of the administered ^{14}C-PCB.
(Ap-volume 0-30 cm in 24 m^2 = 7200 lt x bulk density
1.3 = 9360 Kg. 1 percent C from 9360 Kg = 93,600 g,
multiplied with 100 cpm after nine months = 9,350,000
cpm = 12,448,000 dpm = *ca.* 6 μv = 12 percent). The
largest part of the PCB has been lost under field
conditions by evaporation + leaching + microbial and
photochemical decomposition.

2) Reinbek biotop: Haplumbrept in glacial sand,
12 m^2, treated with 25 μc of ^{14}C-PCB as described before.
Total precipitations during the nine months observation period
were 493.3 mm.
Results from Table IV:

- In the more sandy material the percolation losses seem
to be higher and faster than in the loessic material
of Kleinaltendorf.
- In the umbric epipedon Ah a level amounting to *ca.*
20 percent of the initial specific activity is maintained a rather long time.
- Based on 5 percent organic substance in the Ah horizon,
total C/m^2 amounts to 7500 g C/m^2 (5 percent of 250
Kg till 25 cm depth, bulk density 1.0, 59 percent C
in organic matter). A final activity of 43 dpm/g C
yields 320,000 dpm/m^2 after nine months, equal to *ca.*
8 percent of the administered quantity. Most of the
lost PCB seems to be leached in the sandy material.

Preliminary experiments regarding PCB decomposition
under biotic as well as sterile conditions with or without
UV irradiation were carried out in 500-ml round flasks and
500-ml quartz tubes. Further distinction was made by parallel
tests with maximum water capacity of 80 percent field capacity

Table III

PCB Translocation in the Soil (Biotop Kleinaltendorf)

	25.6 % C	cpm/g C	After ^{14}C-PCB Application				
			6.8 % C	cpm/g C	9.10 % C	cpm/g C	5.2 cpm/g C
Gooseberry Shrub le + br[a]		309		138		81	52
Grass		792		123		43	24
Bare Soil							
Ap 0-30 cm, total soil	0.9	265		192		183	73
2-0.06 mm	1.0	247		260		360	
< 0.06 mm	0.7	343		352		1000	
total humic matter[b]	19.0	264		141		208	
fulvic acid		147		78		118	
humin coal[c]	0.5	167		137		115	
hexane extract		2080		671		622	
extracted soil	0.9	171		171		107	
Al 30-60 cm, total soil	0.6	182		10		10	12
2-0.06 mm	0.6	135		8		13	
< 0.06 mm	0.5	276		11		17	
total humic matter	23.2	183		9		20	
fulvic acid		95		9		15	
humin coal	0.2	90		9		13	
hexane extract		128		60		128	
extracted soil	0.6	126		11		2	
Bt 60-90 cm, total soil	0.4	203		222		10	11
2-0.06 mm	0.45	71		99		18	
< 0.06 mm	0.4	106		255		26	
total humic matter	26.3	148		143		49	
fulvic acid		114		109		33	
humic coal	0.2	85		89		34	
hexane extract		86		323		77	
extracted soil	0.4	189		243		2	

Grass Cover						
Ap 0-30 cm, total soil	1.1	693	1.0	307	353	117
2-0.06 mm	1.3	509	1.2	160	51	
< 0.06 mm	0.8	1312	0.6	291	338	
total humic matter	21.0	752	21.0	287	354	
fulvic acid	0.1	291	0.2	142	132	
humin coal		417		247	154	
hexane extract		1021		889	1054	
extracted soil	1.0	632	1.2	252	296	
Al 30-60 cm, total soil	0.5	62	0.5	12	12	10
2-0.06 mm	0.5	56	0.5	12	12	
< 0.06 mm	0.4	64	0.4	13	12	
total humic matter	18.9	49		13	13	
fulvic acid	0.1	16		27	10	
humin coal		32		9	12	
hexane extract		75		56	18	
extracted soil	0.4	51	0.5	30	12	
Bt 60-90 cm, total soil	0.4	352	0.4	10	12	11
2-0.06 mm	0.5	196		10	8	
< 0.06 mm	0.4	247		12	11	
total humic matter	24.0	216		13	10	
fulvic acid	0.2	67		16	9	
humin coal		144		10	5	
hexane extract		442		121	27	
extracted soil	0.5	230		9	10	
Rain Worms 30-90 cm					62	

[a] le + br = leaves + branches.
[b] Total humic matter = humic acid + humine.
[c] Humic coal = Organic residue after humic acid and humine extraction.

Table IV
[14]C–PCB Activities of the Horizons of Biotop Reinbek
Within Nine Months Observation Period

Horizon (cm)	dpm/gC in the Samples of Eight Sampling Procedures							
	1975						1976	
	5/28	6/26	8/26	10/13	11/13	12/15	1/14	2/19
Ah 0–25	206	86	80	71	64	52	41	43
AhBv 25–45	143	65	23	23	14	8	–	–
Bv 45–65	141	21	16	11	–	–	6	–
BvCv 65–85	48	24	6	6	10	–	–	–

state of moisture in the soil. Sterilization was maintained
after autoclaving by addition of 0.5% Na-ethylmercurithio-
salicylate, whose effect was ascertained by plate tests. PCB
degradation has been assessed on the basis of [14]CO_2 evolved,
which was transferred by means of a vacuum line with Toepler
pump in a 1-liter ionization chamber vibrating reed electrom-
eter system. For refilling of the reaction vessels sterile
air, passed through a Dovex hollow fiber beaker, gas perme-
ator, b/HFG-1 was used.

According to Table V PCB degradation is considerably
higher at 80 percent field capacity than at maximum water
capacity. In the sterile tests the PCB decomposition is
lowered but definitely existing. The sums of eight measurements from June 12 till October 29 are revealed in Table V.
Additional PCB degradation under influence of UV irradiation
is revealed by Table VI. Under maximum water capacity state
of moisture, the photochemical PCB decay is low. At 80 percent field capacity the [14]CO_2-release is slightly enhanced.
in the case of the biotic, and considerably increased in the
sterile sample due to UV irradiation.

SUMMARY

Adsorption studies with ^{14}C-labelled PCB on montmorillonite humic and fulvic acid indicate C-type linear adsorption
isotherms. In central European Hapludalf, with more illitic
clay, costume adsorption occurs rather selectively in contact
with organic rotting products, as thin section autoradiography
reveals. This observation can be confirmed also from translocation studies in two small biotops with loessic Hapludalf
and glacial sandy Haplumbrept soils. After eight to nine
months PCB was left in the organic matter containing A-horizons
(epipedons) only, at an extent < 13 percent of the initially
administered amount.

Table V
PCB Decomposition under Biotic and Sterile Conditions,
at Maximum Water Capacity and
80 Percent Field Capacity Moisture Level

Backgrounds of Ionization Chambers (μc)	Maximum Water Capacity		80% Field Capacity Level	
	Biotic(1) (μc)	Sterile(2) (μc)	Biotic(3) (μc)	Sterile(4) (μc)
166.2×10^{-5}	$\begin{array}{c} 3160 \times 10^{-5} \\ -166 \\ \hline 299 \times 10^{-5} \end{array}$	$\begin{array}{c} 2044 \times 10^{-5} \\ -166 \\ \hline 1878 \times 10^{-5} \end{array}$	$\begin{array}{c} 7800 \times 10^{-5} \\ -166 \\ \hline 7634 \times 10^{-5} \end{array}$	$\begin{array}{c} 3700 \times 10^{-5} \\ -166 \\ \hline 3534 \times 10^{-5} \end{array}$
Average of 8 Measurements	374	234	954	442
% Relative to (3)	39.2	24.5	100	46.3
(2) in % of (1)	62.73%			
(4) in % of (3)			46.29%	

Table VI
PCB Decomposition under Biotic and Sterile Conditions,
at Maximum Water Capacity and 80 Percent
Field Capacity Moisture Level,
All under UV Irradiation

	Maximum Water Capacity		80% Field Capacity Level	
	Biotic(1) (μc)	Sterile(2) (μc)	Biotic(3) (μc)	Sterile(4) (μc)
Sum of 4 or 5 Measurements (-Background)	136×10^{-5}	28×10^{-5}	8871×10^{-5}	4358×10^{-5}
Divided by Number of Repetitions	34	7	1774	1089
% Relative to (3)	1.9	0.4	100	61.4
(2) in % of (1)	20.59%			
(4) in % of (3)			61.39%	

ACKNOWLEDGMENTS

We thank Dr. W. Klein of the Institut fur Umweltchemie, Schloß Birlinghofen for providing the ^{14}C-PCB samples. E. Kruse and A. Lay have been very helpful in soil fractionation and benzene synthesis. The Federal Department of Research and Technology as well as the Bayer A.G. provided the funds. B. K. G. Theng acknowledges the Alexander von Humboldt Stiftung for a fellowship enabling him to participate in this study.

REFERENCES

Baxter, R. A., P. E. Gilbert, R. A. Lidgett, R. H. Mainprize and H. A. Vodden. "The Degradation of Polychlorinated Biphenyls by Microorganisms," *The Science of the Total Environment* 4:58-61 (1975).

Bykov, V. T., V. G. Gerasimova, G. A. Krizhanenko and V. P. Shvetsov. "Chromatographic Study of the Adsorption of Some Polar and Nonpolar Compounds on Kaolinite and Montmorillonite," *Kolloid. Zh.* 38:542-544 (1976).

Edwards, R. "Polychlorobiphenyls, Their Occurrence and Significance," *Chem. Ind.* (London) 47:1340-1348 (1971).

Eltantawy, I. M. and P. W. Arnold. "Adsorption of *n*-Alkanes by Wyoming Montmorillonite," *Nature, Phys. Sci.* 237:123-125 (1972).

Fischer, H. A. and G. Werner. *Autoradiographic* (Berlin: De Gruyter, 1971).

Giles, C. H., T. H. MacEwan, S. M. Nakhwa and D. Smith. "Studies in Adsorption. Part XI. A System of Classification of Solution Adsorption Isotherms and its Use in Diagnosis of Adsorption Mechanisms and in Measurement of Specific Surface Areas of Solids," *J. Chem. Soc.* 3 73-3993 (1960).

Gustafson, C. G. "PCB's - Prevalent and Persistent. Intensified Research is Needed to Minimize their Dangers," *Environ. Sci. Technol.* 4:814-819 (1970).

Hammond, A. L. "Chemical Pollution: Polychlorinated Biphenyls," *Science* 175:155-156 (1972).

Haque, R. and D. Schmedding. "Studies on the Adsorption of Selected Polychlorinated Biphenyl Isomers on Several Surfaces," *J. Environ. Sci. Health* B11:129-137 (1976).

Haque, R., D. W. Schmedding and V. H. Freed. "Aqueous Solubility, Adsorption and Vapor Behavior of Polychlorinated Biphenyl Aroclor 1254," *Environ. Sci. Technol.* 8:139-142 (1974).

Harvey, G. H. and W. G. Steinhauer. "Biochemistry of PCB and DDT in the North Atlantic," *Proceedings of the 2nd International Conference on Environmental Biogeochemistry,* Hamilton, Canada (1975) pp. 203-225.

Jensen, S., A. G. Johnels, M. Olsson and G. Otterlind. "DDT and PCB in Marine Animals from Swedish Waters," *Nature* 224:247-250 (1969).

Kaiser, K. L. E. and P. T. S. Wong. "Bacterial Degradation of Polychlorinated Biphenyls. I. Identification of Some Metabolic Products from Aroclor 1242," *Bull. Environ. Contamination and Toxicol.* 11:291-296 (1974).

Kerpen, W. and H. W. Scharpenseel. "Movement of Ions and Colloids in Undisturbed Soil and Parent Rock Material Columns," *Proceedings of the International Conference on Isotope and Radiation Techniques in Soil Physics and Irrigation Studies,* Istanbul (1967), pp. 213-216.

Koeman, J. H., M. C. Tennoever De Brauw and R. H. De Vos. "Chlorinated Biphenyls in Fish, Mussels and Birds from the River Rhine and the Netherlands Coastal Area," *Nature* 221:1126-1128 (1969).

National Conference on Polychlorinated Biphenyls Proceedings EPA-560/6-75-004. (Washington, D.C.: Environmental Protection Agency, Office of Toxic Substances, 1975).

Nelson, N., P. B. Hammond, I. C. T. Nisbet, A. F. Sarofim and W. H. Drury. "Polychlorinated Biphenyls. Environmental Impact," *Environ. Res.* 5:249-362 (1972).

Peakall, D. B. and J. L. Lincer. "Polychlorinated Biphenyls. Another Long-Life Widespread Chemical in the Environment," *Bioscience* 20:958-964 (1970).

Risebrough, R. W., P. Rieche, D. B. Peakall, S. G. Herman and M. N. Kriven. "Polychlorinated Biphenyls in the Global Ecosystem," *Nature* 220:1098-1102 (1968).

Risebrough, R. W., V. Vreeland, G. R. Harvey, H. P. Miklas and G. M. Carmignani. "PCB Residues in Atlantic Zooplanktonn" *Bull. Environ. Contamination Toxicol.* 8:345-355 (1972).

Scharpenseel, H. W. and H. Beckmann. "Untersuchungen zur Kohlendioxidentbindung des Bodens," *A. Pflanzenernähr. Düng. Bodenk.* 104:110-118 (1964).

Scharpenseel, H. W. and F. Pietig. "Einfache Boden- und Wasserdatierung durch Messung der ^{14}C-oder Tritiumkonzentration," *Geoderma* 2:273-289 (1969).

Scharpenseel, H. W., C. Ronzani and F. Pietig. "Comparative Age Determination on Different Humic-Matter Fractions," *Proceedings of a Symposium on Isotopes and Radiation in Soil Organic-Matter Studies*, Vienna (1960) pp. 67-73.

Scharpenseel, H. W., B. K. G. Theng and S. Stephan. "Infiltration und Translokation von polychlorierten Biphenylen in natürlich gelagerten Bodenprofilen; biotischer und abiotischer Abbau. 1. Adsorption und Einbau polychlorierter Biphenyle (PCB) im Boden," *Z. Pflanzenernähr Bodenk.* (in press, 1977a).

Scharpenseel, H. W., B. K. G. Theng and S. Stephan. "Infiltration und Translokation von polychlorierten biotischer und abiotischer Abbau. 2. Verteilung und Abbau polychlorierter Biphenyle (PCB) im Boden," *A. Pflanzenernah Boden* (in press, 1977b).

Solly, S. R. B. and V. Shanks. "Polychlorinated Biphenyls and Organochlorine Pesticides in Human Fat in New Zealand," *N. A. J. Sci.* 17:535-544 (1974).

Stephan, S. "Gefriertrocknung und andere bei der Herstellung von Dodendünnschliffen benutzbare Trocknungsverfahren," *Z. Pflanzenernähr. Bodenk.* 123:131-140 (1969).

Theng, B. K. G. "Adsorption of Ammonium and Some Primary *n*-Alkylammonium Cations by Soil Allophane," *Nature* 238: 150-151 (1972).

Theng, B. K. G. and H. W. Scharpenseel. "The Adsorption of ^{14}C-Labelled Humic Acid by Montmorillonite," *Proceedings of the International Clay Conference*, Mexico City, (1975). pp. 643-653.

Tucker, E. S., W. J. Litschgi and W. M. Mees. "Migration of Polychlorinated Biphenyls in Soil Induced by Percolating Water," *Bull. of Environ. Contamination Toxicol.* 13:86-93 (1975).

Weil, L., G. Dure and K. E. Quentin. "Adsorption von chlorierten Kohlenwasserstoffen an organischen Wassertrübstoffen und an Böden," *Z. Wasser- Abwasser-Forsch* 6:107-112 (1973).

Weil, L., K. E. Quentin and S. Jarrar. "Zur Analytik der Pestizide im Wasser," *Schriftenreihe Verein fur Wasser-Boden- und Luft-Hygiene, Berlin-Dahlem* 37:77-84 (1972).

Yobs, A. R. "Levels of Polychlorinated Biphenyls in Adipose Tissue of the General Population of the Nation," *Environ. Health Perspectives* 1:79-81 (1972).

CHAPTER 51

REACTIONS OF HUMIC SUBSTANCES
WITH MINERALS IN THE SOIL ENVIRONMENT

M. SCHNITZER

Soil Research Institute
Agriculture Canada
Ottawa, Ontario, Canada

INTRODUCTION

Humic substances, the major organic constituents of
soils and waters, have the ability to interact with metal ions,
metal oxides and hydroxides and more complex soil minerals
to form metal-organic associations of widely differing chemical and biological stabilities and characteristics. Humic
materials contain, per unit weight, relatively large numbers
of oxygen-containing functional groups (CO_2H, phenolic OH
and C=O), through which they can attack and degrade soil
minerals by complexing and dissolving metals and transporting
these within soils and waters (Schnitzer and Skinner, 1963;
Rosell and Babcock, 1968; Baker, 1973; Schnitzer and Kodama,
1976; Singer and Navrot, 1976).
    While relatively little is known about the chemistry
of the metal-humic interaction products, it is likely that
these associations affect many reactions in soils and waters.
One potentially important reaction that soil and water scien-
tists have so far tended to underestimate is the known ability
of inorganic surfaces to catalyze organic reactions. Thus,
in soils and waters, where relatively large amounts of humic
materials are closely associated with mineral surfaces, catalytic reactions may be important in the synthesis, alteration
and degradation of humic materials. Conversely, humic substances may favor the crystallization of organically complexed
inorganics under unusually favorable conditions.

Characteristics of Humic Substances

Humic substances are amorphous, dark-colored, hydro-philic, acidic, predominantly aromatic, chemically complex organic substances that range in molecular weight from a few hundred to several thousand.

Based on their solubility in alkali and acid, humic substances are usually partitioned into three main fractions: (a) humic acid (HA), which is soluble in dilute alkali but is precipitated by acidification of the alkaline extract; (b) fulvic acid (FA), which is that humic fraction remaining in solution when the alkaline extract is acidified; that is, it is soluble in both dilute alkali and in dilute acid; and (c) humin, which is that humic fraction that cannot be extracted from the soil or sediment by dilute base and acid. From analytical data published in the literature (Schnitzer and Khan, 1972) it becomes apparent that structurally the three humic fractions are similar, but differ in molecular weight, ultimate analysis, and functional group content, with FA having a lower molecular weight but higher content of oxygen-containing functional groups (CO_2H, OH, C=O) per unit weight than the other two humic fractions.

Of the three principal humic fractions only FA is completely soluble in water at all pH values known to prevail in soils and waters, whereas HA is soluble in water at pH > 6.5 only, and humin is insoluble in water at pH's normally found in soils and waters.

Stability Constants of Metal-HA and -FA Complexes

It may be appropriate to refer at this point to measurements of metal-HA and metal-FA stability constants that have been done in recent years in a number of laboratories (Schnitzer, 1971; Schnitzer and Khan, 1972; Stevenson and Ardakani, 1972; Gamble and Schnitzer, 1973; Jellinek, 1974; Flaig *et al.*, 1975). For example, stability constants, expressed as log K, for metal-FA complexes were determined by Schnitzer and Hansen (1970) by the method of continuous variations and by the ion-exchange equilibrium method. These data are shown in Table I. The stability constants measured by the two methods were in good agreement with each other, increased with increase in pH, but decreased with increase in ionic strength. Of all metals investigated, Fe^{3+} formed the most stable complex with FA. The order of stabilities at pH 3.0 was: $Fe^{3+} > Al^{3+} > Cu^{2+} > Ni^{2+} > Co^{2+} > Pb^{2+} > Ca^{2+} > Zn^{2+} > Mn^{2+} > Mg^{2+}$. At pH 5.0, stability constants of Ni-FA and Co-FA complexes were slightly higher than those of the Cu-FA complex. The stability constants shown in Table I are considerably lower than those for complexes formed between the same metal ions and synthetic complexing agents such as

Table I
Stability Constants of Metal-FA Complexes[a]

| Metal | pH 3.0 | | pH 5.0 | |
	CV[b]	IE[c]	CV	IE
Cu^{2+}	3.3	3.3	4.0	4.0
Ni^{2+}	3.1	3.2	4.2	4.2
Co^{2+}	2.9	2.8	4.2	4.1
Pb^{2+}	2.6	2.7	4.1	4.0
Ca^{2+}	2.6	2.7	3.4	3.3
Zn^{2+}	2.4	2.2	3.7	3.6
Mn^{2+}	2.1	2.2	3.7	3.7
Mg^{2+}	1.9	1.9	2.2	2.1
Fe^{3+}	6.1[d]	-	-	-
Al^{3+}	3.7[e]	3.7	-	-

[a] Schnitzer and Hansen, 1970.
[b] CV = Method of continuous variations.
[c] IE = Ion-exchange equilibrium method.
[d] Determined at pH 1.70.
[e] Determined at pH 2.35.

EDTA. This may mean that metals complexed by FA should be more readily available to plant roots, microbes and small animals than when sequestered by EDTA or similar reagents.

Mechanisms of Principal Types of Interactions Between HA and FA and Soil Minerals

From the data available so far there are three principal reaction mechanisms governing HA- and FA-mineral interactions that can be postulated. These will be discussed in the following sections.

Dissolution of Minerals

In view of their ability to complex mono-, di-, tri- and tetravalent metal ions, FA's and, at pH > 6.5, also HA's can attack and degrade minerals to form water-soluble and water-insoluble metal complexes. If metal/HA or metal/FA ratios are low, the complexes are water-soluble, but if the ratios are high, the metal-HA or -FA complexes become water-insoluble (Schnitzer and Khan, 1972). Strong solvent activity has been reported for HA's for hematite, pyrolusite, feldspar, biotite, enstatite, actinolite and epidote (Baker, 1973). Some of Baker's data are shown in Table II. Mn^{3+}- and Mn^{4+}-oxides and Mn^{2+}-hydroxides (Rosell and Babcock, 1968) and basalt (Singer and Navrot, 1976) are also gradually dissolved. Similarly, FA's have been shown to be efficient for dissolving metals from goethite, gibbsite and soils (Schnitzer and Skinner, 1963), chlorites (Table III, Kodama and Schnitzer, 1973), and micas (Table IV, Schnitzer and Kodama, 1976).

Table II
Reactions of H_2O and HA (pH 3.4) with Silicates[a]

Mineral	Element Extracted	μg Extracted in 5 Days by H_2O	0.1% HA
Feldspar	Na	340	1260
	K	190	640
	Al	10	220
Biotite	K	60	100
	Mg	20	90
	Fe	5	240
	Al	10	300
Actinolite	Fe	5	960
	Ca	750	2900
	Si	5	40

[a]Baker, 1973.

Table III
Dissolution of Metals (Fe, Al, Mg) from Chlorites
by 0.2% FA Solution and by H_2O after 360 hr
of Shaking at pH 2.5[a]

Type of Mineral	% of Sample Dissolved by H_2O	0.2% FA
Leuchtenbergite (Fe-poor)	2	4
Thuringite (Fe-rich)	6	26

[a]Kodama and Schnitzer, 1973.

Table IV
Dissolution of Micas by 0.2% FA (pH 2.5)[a]

Mineral	Element Dissolved	% Dissolved after 710 hr
Biotite	Fe	11.5
(Fe-rich)	Al	14.5
	Mg	17.0
	K	18.5
	Si	14.0
Phlogopite	Al	8.0
(Fe-poor)	Mg	8.5
	K	8.0
	Si	9.0
Muscovite	Al	2.3
(Fe-poor)	K	1.8
	Si	0.6

[a]Schnitzer and Kodama, 1976.

IR and ESR analyses of water-soluble and water-insoluble metal-HA and -FA complexes show that humic CO_2H and phenolic OH groups in combination or CO_2H groups only are involved in metal complexing (see Figure 1, a and b). A point of considerable interest is illustrated by the data in Tables III and IV, which show that minerals rich in iron are most susceptible to attack by humic materials (Schnitzer and Kodama, 1976). The high affinity of HA or FA for iron

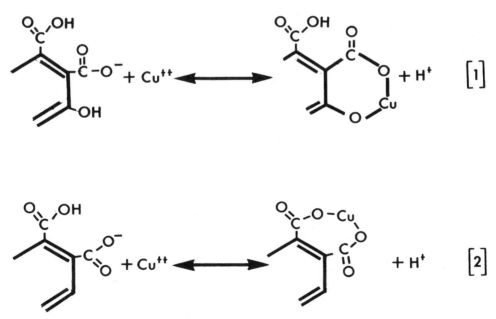

Figure 1. Reactions between FA and Cu^{++}.

and the low crystal field stabilization energies of Fe^{3+} and Fe^{2+} provide possible explanations for these findings. The complexing of iron by humic materials appears to have an adverse effect on the structural stability of Fe-rich minerals, as along with Fe, other major constituent elements such as Mg, Al, K and Si are also more readily released and brought into solution. Recently, Senesi *et al.*, (1977) have combined ESR and Mössbauer spectroscopy with chemical treatments to obtain information on oxidation states and site symmetries of iron bound by HA and FA. At least two, possibly three, different binding sites for iron were found to occur in humic materials: (a) Fe^{3+} was strongly bound and protected by tetrahedral and/or octahedral coordination; this form of iron exhibited considerable resistance to complexation by known sequestering agents and reduction; and (b) Fe^{3+} was adsorbed on external surfaces of humic materials, weakly bound octahedrally, and easily complexed and reduced. These observations were made on laboratory-prepared Fe-HA and Fe-FA complexes and on a Fe-pan taken from a soil rich in organic matter.

In a recent study on water-soluble Mn^{2+}-FA complexes, Gamble *et al.*, (1976) noted that $1/2Mn^{2+}$ and K^+ were complexed to FA by similar electrostatic binding, so that in aqueous solutions Mn^{2+} is at least partially bound electrostatically as $Mn(OH_2)_6^{2+}$, with the FA donor groups in outer sphere complexing sites. It is likely that the outer sphere complex is unsymmetrical. This situation is somewhat similar to that postulated for reactions between clays and organic molecules. Yariv *et al.*, (1966) stress the formation of "water bridges," linking polar molecules to exchangeable metal cations through water molecules in the primary hydration shell by a mechanism such as the following:

$$
M^{n+} \quad
\begin{matrix} H \\ | \\ O \end{matrix}
- H \ldots\ldots O =
\begin{matrix} O^- \\ | \\ C \\ | \\ R \end{matrix}
$$

where M^{n+} is the metal cation and R is the rest of the molecule. The bond between O and H is a hydrogen bond, and this type of bond is of special importance when the cation has a high solvation energy and so retains its primary hydration shell. In our case $RCOO^-$ stands for partially ionized FA and M^{n+} for Mn^{2+}, with six H_2O molecules in its primary hydration shell. Thus, these data suggest an outer sphere electrostatic structure for the complex, with both electrostatic attraction and hydrogen bonding occurring simultaneously. Other transition metals may react with FA by similar mechanisms to form water-soluble complexes.

Adsorption on External Mineral Surfaces

The extent of adsorption of humic materials on mineral surfaces depends on the physical and chemical characteristics of the surface, the pH of the system and its water content. One can visualize the formation of a wide range of mineral-humic associations, involving chemical bonding with widely differing strengths. Another mechanism of considerable importance is hydrogen-bonding, the occurrence of which is clearly indicated in IR spectra of HA- and FA-mineral complexes (Schnitzer and Khan, 1972). These reactions are likely to involve H and O of CO_2H and OH groups in HA and FA and O and H on mineral surfaces and edges. It is probably, as has been mentioned above, that cations with high solvation energies on mineral surfaces react via water-bridges with HA- and FA-functional groups. Van der Waals' forces may also contribute to the adsorption of humic substances on mineral surfaces.

The importance of surface geometry and surface chemistry has been pointed out by a number of workers. Inoue and Wada (1971) studied reactions between humified clover and imogolite and postulated the following two mechanisms: (1) incorporation of CO_2H groups into the coordination shell of Al atoms, and (2) relatively weak bonding by H-linkages and van der Waals' forces. The two adsorption mechanisms were thought to assist each other and the number of bonds formed was considered to depend on the orientation of the humic molecules on the clay. Kodama and Schnitzer (1974) report high adsorption of FA on sepiolite surfaces. Sepiolite has a channel-like structure formed by the joining of edges of long and slender tall structures. In untreated sepiolite the channels are occupied by bound and/or zeolitic water, which apparently can be displaced by undissociated FA. IR spectra of FA-sepiolite interaction products suggest the formation in the channels of COO^- groups, which are linked to Mg^{2+} at edges that have been exposed by the displacement of water by FA.

Adsorption in Clay Interlayers

The evidence currently available shows that the inter-layer adsorption by expanding clay minerals of humic materials is pH-dependent, being greatest at low pH, and no longer occurring at pH > 5.0 (Schnitzer and Kodama, 1966). Adsorbed FA cannot be displaced from clay interlayers by leaching with 1 N NaCl, and an inflection occurs in the adsorption-pH curve near the pH corresponding to the pH of the acid species of FA (Schnitzer and Kodama, 1966). On the basis of these criteria, the adsorption could be classified as a "ligand-exchange" reaction (Greenland, 1971). In this type of adsorption the anion is thought to penetrate the coordination shell of the dominant cation in the clay and displace water coordinated to the dominant cation in the clay interlayer.

The ease with which water can be displaced will depend on the affinity for water of the dominant cation with which the clay is saturated and also on the degree of dissociation of the FA. Since the latter is very low at low pH, inter-layer adsorption of FA is greatest at low pH levels as Schnitzer and Kodama (1966) have observed. Concurrently the FA can dissolve a proportion of the dominant cation in the clay by forming a soluble complex and replacing the removed cation by H^+. If this process continues over long periods of time, the FA will eventually degrade the clay structure.

REFERENCES

Baker, W. E. "Role of Humic Acids from Tasmanian Podzolic Soils in Mineral Degradation and Metal Mobilization, *Geochim. Cosmochim. Acta* 37:269-281 (1973).

Flaig, W., H. Beutelspacher and E. Rietz. "Chemical Composition and Physical Properties of Humic Substances," in *Soil Components*, Vol. I., J. E. Gieseking, ed. (New York: Springer Verlag, 1975), pp. 1-211.

Gamble, D. S. and M. Schnitzer. "The Chemistry of Fulvic Acid and Its Reactions with Metal Ions," in *Trace Metals and Metal-Organic Interactions in Natural Waters*, P. C. Singer, ed. (Ann Arbor, Michigan: Ann Arbor Science Press, 1973), pp. 265-302.

Gamble, D. S., M. Schnitzer and D. S. Skinner. "Mn^{II}-Fulvic Acid Complexing Equilibrium Measurements by Electron Spin Resonance Spectrometry," *Can. J. Soil Sci.* (in press).

Greenland, D. J. "Interactions Between Humic and Fulvic Acids and Clays," *Soil Sci.* 111:34-41 (1971).

Inoue, T. and K. Wada. "Reactions Between Humified Clover Extract and Imogolite as a Model of Humus-Clay Interaction. Part II," *Clay Sci.* 4:61-70 (1971).

Jellinek, H. H. G. *Soil Organics. I. Complexation of Heavy Metals* Special Report 212 (Hannover, Hew Hampshire: Cold Regions Research and Engineering Laboratory, 197), pp. 1-51.

Kodama, H. and M. Schnitzer. "Dissolution of Chlorite Minerals by Fulvic Acid," *Can. J. Soil Sci.* 53:240-243 (1973).

Kodama, H. and M. Schnitzer. "Adsorption of Fulvic Acid by Nonexpending Clay Minerals," *10th Intern. Cong. Soil Sci., Proc. II* Moscow, USSR (1974), pp. 51-56.

Rosell, R. A. and K. L. Babcock. "Precipitated Manganese Isotopically Exchanged with ^{54}Mn and Chelated by Soil Organic Matter," in *Isotopes and Radiation in Soil Organic Matter Studies* (Vienna: International Atomic Energy Agency, 1968), pp. 453-469.

Schnitzer, M. "Metal-Organic Matter Interactions in Soils and Waters," in *Organic Compounds in Aquatic Environments* S. D. Faust and J. D. Hunter, eds., (New York: Marcel Dekker, 1971), pp. 297-315.

Schnitzer, M. and E. H. Hansen. "Organo-Metallic Interactions in Soils: 8. An Evaluation of Methods for the Determination of Stability Constants of Metal-Fulvic Acid Complexes," *Soil Sci.* 109:333-340 (1970).

Schnitzer, M. and S. U. Khan. *Humic Substances in the Environment* (New York: Marcel Dekker, 1972).

Schnitzer, M. and H. Kodama. "Montmorillonite: Effect of pH on Its Adsorption of a Soil Humic Compound," *Science* 153:70-71 (1966).

Schnitzer, M. and H. Kodama. "The Dissolution of Micas by Fulvic Acid," *Geoderma* 15:381-391 (1976).

Schnitzer, M. and S. I. M. Skinner. "Organo-Metallic Interactions in Soils. 2. Reactions Between Different Forms of Iron and Aluminum and the Organic Matter of a Podzol Bh Horizon," *Soil Sci.* 96:181-186 (1963).

Senesi, N., S. M. Griffith, M. Schnitzer and M. G. Townsend. "Binding of Fe^{3+} by Humic Materials," *Geochim. Cosmochim. Acta* (submitted, 1977).

Singer, A. and J. Navrot. "Extraction of Metals from Basalt by Humic Acids," *Nature* 262:479-480 (1976).

Stevenson, F. J. and M. S. Ardakani. "Organic Matter Reactions Involving Micronutrients in Soils," in *Micronutrients in Agriculture* J. J. Mortvedt, P. M. Giardano and W. L. Lindsay, eds. (Madison, Wisconsin: Soil Sci. Soc. Am., 1972), pp. 79-114.

Yariv, S., J. D. Russel and V. C. Farmer. "Infrared Study of the Adsorption of Benzoic Acid and Nitrobenzene in Montmorillonite," *Israel J. Chem.* 4:201-213 (1966).

CHAPTER 52

THE PROTECTIVE EFFECT OF BUFFERS ON THE INACTIVATION OF UREASE SORBED TO MONTMORILLONITE AND KAOLINITE CLAY SOILS AND MINERALS

H. MAKBOUL
J. C. G. OTTOW

Institut für Bodenkunde und Standortslehre
Universität Hohenheim (LH)
7 Stuttgart, Hohenheim, Federal Republic of Germany

INTRODUCTION

Pelozolic soils are those terrestrial habitats characterized by a high clay content (30-60 percent). Such clays not only govern gross ecological properties such as swelling-shrinking phenomena, water movement or interlattice fixation of nutrients, but may determine particularly biochemical processes that occur essentially in the microhabitat of the clay surface. Thus, growth of various organisms (Martin et al., 1976), their metabolism (Kunc and Stotzky, 1974), the turnover rate or organic and inorganic substrates (Stotzky, 1972) as well as the activity of various exo-enzymes (Durand, 1965) are affected by the clays in different ways and to quite varying extents. Soil enzymes, once sorbed to the clay surfaces, become stabilized and protected against hydrolysis (Burns et al., 1972), though their inactivation may continue with increasing time (Skujins, 1967). These inactivation processes should depend on the physicochemical properties and the amount of clay as well as on the buffering action of the soil solution in question.

In this chapter, the effect of montmorillonite and kaolinite clay types on the inactivation rate of urease is reported. Further, the influence of various buffer solutions on desorption or reactivation of urease, either sorbed to montmorillonite and kaolinite or present in naturally occurring clay soils, is presented.

MATERIALS AND METHODS

Pelozolic Soils

The clay material used in these experiments was
obtained from two pelozolic soil types; one, located near
Metzingen, Germany, contains 25-30 percent kaolinite, the
other, at Wurmlingen, about 30 percent montmorillonite with
traces of illite. Samples were collected from the subsoil
at 5-40 cm (Bv-horizon) and 15-50 cm (Bva), respectively,
air-dried, ground and sieved to pass a 2-mm screen. The
Bv-material was acid (pH KCl = 3.7) and contained 0.3% N
and 6.7% C, whereas the Bva-clay revealed a pH KCl of 6.6
showing a total N of 0.2% and a C-content of 3.1%.

Clay Minerals

Commercial Ca-montmorillonite and Ca-kaolinite were
obtained from the Geisenheimer Kaolinwerke, 6222 Geisenheim,
West Germany. The cation exchange capacities (CEC) of these
products were 0.72 and 0.26 meq/g of montmorillonite and
kaolinite, respectively.

Urease Stock Solution

Crystalline Jack-bean urease (article no. 37799) was
purchased from Serva, Heidelberg, West Germany. To obtain
the stock solutions, 100 mg of urease was dissolved in
250 ml double distilled water. One ml of this solution
produced 3792 µg NH_4/2 hr from urea as determined according
to Tabatabai and Bremner (1972).

Buffer Solutions

The buffers used to extract the enzyme from the clay
minerals or from the two soils used were as follows:

1. THAM: 0.05 M tris (hydroxymethyl) aminomethan
   adjusted either to pH 7.6 or 9.0.
2. $Na_4P_2O_7$: 0.025 M (at pH 7.6 and 9.0).
3. Mixture from 0.05 M THAM and 0.025 $Na_4P_2O_7$ (1:1)
   corrected for pH 7.6 and 9.0, respectively.

Determination of Urease in Sediment and Extract

One g clay mineral was weighed into a 25-ml centrifuge-
glass, and mixed gently with 0.5 ml toluene. The tube was
swirled a few seconds and left for 15 min before adding 1
ml of the urease stock solution. After 15 min (room temperature) 10 ml buffer solution (0.05 M THAM at pH 9.0) or

distilled water were added, the glass stoppered, homogenized
and allowed to stand (in darkness at room temperature) for
the desired sorption time (0, 1, 2, 15, 24 and 48 hr). After
centrifugation (30 min at 6000 rpm) the extract was decanted
into a 50-ml volumetric flask, and both the clay mineral
(sediment) and extract were assayed for urease activity,
following the method described by Tabatabai and Bremner
(1972). Controls were run without clay minerals. Urease
activity was expressed in NH_4-released ($\mu g/g$ clay-sediment/2
hr or in $\mu g/5$ g soil/2 hr). For the extract the urease
activity was given in μg $NH_4/10$ ml solution/2 hr.

Effect of Buffer on Native Soil Urease

Five g soil (Bv- or Bva-material) was placed in a
50-ml volumetric flask, mixed with 0.5 ml toluene, swirled
and allowed to stand for 15 min. Each sample was mixed with
10 ml buffer solution (0.05 M THAM; 0.025 M $Na_4P_2O_7$ or mixed
1:1) of different pH values (7.6 and 9.0), gently shaken and
allowed to stand at room temperature for 2, 24 and 72 hr,
respectively, before determining the urease activity. The
values listed are the average of three determinations.

RESULTS

Inactivation Rate of Urease by Clay Minerals

Table I presents the inactivation of urease by montmorillonite and kaolinite. The following results are obvious.
First, urease activity in the sediment plus extract remained
significantly higher in the samples treated with THAM than in
those incubated with water. This is true both with montmorillonite and kaolinite. Second, if the urease activity of the
whole sample (sediment plus extract) is compared to the combined activity of sediment and its extract, considerable
differences are noted. The extract of the montmorillonite
clay was even free from urease activity, suggesting a complete
fixation by the clay. Kaolinite extract gave only very weak
NH_4 determinations. In the case of water, the extracts of
both montmorillonite and kaolinite failed to give any urease
activity. Third, with increasing time, the urease activity
decreased rapidly, both in the sediment alone as well as in
the whole sample (sediment plus extract). This inactivation
was particularly pronounced during the first two hours of
incubation (80 and 74 percent for montmorillonite and kaolinite,
respectively). The development of the inactivation rates in
the sediment plus extract was quite similar for montmorillonite
and kaolinite. This is remarkable, as montmorillonite had a
much higher cation exchange capacity than the kaolinite
(approximately three times). Obviously, the inactivation of
sorbed urease is not directly related to the cation exchange
capacity.

Table I

The Effect of THAM-Buffer (0.05 *M*), Water and Incubation Time on the Urease Activity[a] in the Extract and/or Sediments (Montmorillonite or Kaolinite) from Model Experiments

Time (h)	Montmorillonite (CEC = 0.72 meq/g)				Kaolinite (CEC = 0.26 meq/g)			
	in THAM				in THAM			
	Sed. + Ex.	Sed.	Ex.[b]	Sed. + Ex.	Sed. + Ex.	Sed.	Ex.	Sed. + Ex.
0	1360 (64%)[c]	805 (79%)	0	120 (95%)	1414 (63%)	1115 (71%)	39	952 (75%)
1	1288 (66%)	754 (80%)	0	114 (97%)	1400 (63%)	786 (79%)	39	809 (79%)
2	753 (80%)	418 (89%)	0	107 (97%)	1005 (74%)	282 (93%)	49	726 (81%)
15	759 (80%)	378 (90%)	0	72 (98%)	970 (75%)	126 (97%)	30	284 (93%)
24	571 (85%)	375 (90%)	0	63 (98%)	826 (78%)	82 (98%)	26	282 (93%)
48	516 (86%)	373 (90%)	0	35 (99%)	262 (93%)	0	0	209 (97%)

[a]Expressed in µg NH$_4$-released/g sediment or 10 ml extract. The control (1 ml urease stock solution without clay minerals) liberated 3792 µg NH$_4$/2 hr.

[b]Either the H$_2$O-extract or its sediment alone failed to give any urease activity.

[c]The percentages in parentheses represent the inactivation rate of urease referred to the control.

Effect of Buffer and pH on Urease Activity of Pelozolic Soil

    In Figure 1 the effect of THAM, pyrophosphate and
THAM-pyrophosphate buffer at pH 7.6 as well as at pH 9.0 on
the urease activity of the Bv (kaolinite) clay sample is given.
The same effect on the native urease activity of the Bva

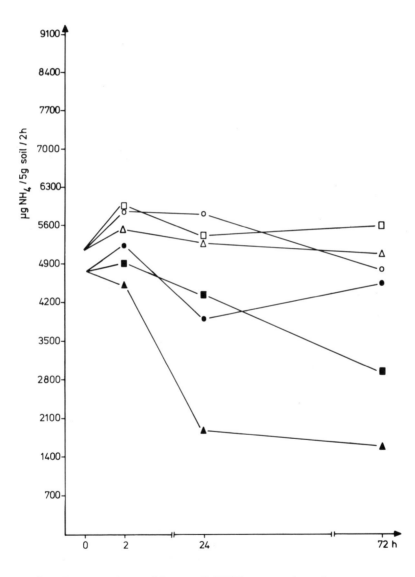

Figure 1. Protective effect of THAM, pyrophosphate or
    THAM-pyrophosphate buffer at pH 7.6 (black) and pH 9.0 (open)
    on urease activity of a naturally occurring kaolinite-con-
    taining soil (Bv, Metzingen). THAM; Na-pyrophosphate; THAM +
    pyrophosphate.

(montmorillonite + illite)-material is presented in Figure 2. The following overall conclusions may be drawn from both graphs.   First, the protective effect of buffers on the

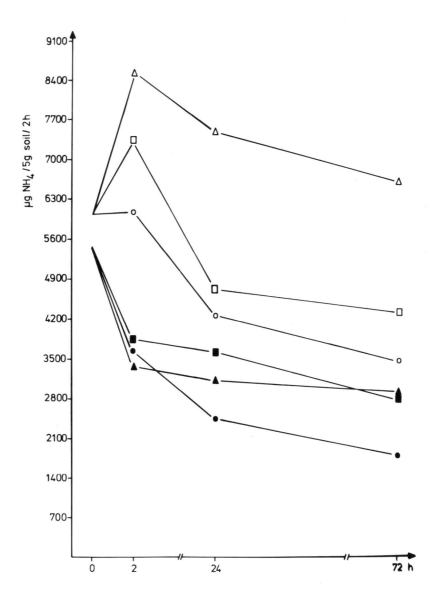

Figure 2.   Protective effect of THAM-, pyrophosphate- or THAM-pyrophosphate buffer at pH 7.6 (black) and pH 9.0 (open) on urease activity of a naturally occurring montmoril-lonite-containing soil (Bva, Wurmlingen).   THAM; Na-pyrophosphate; THAM + pyrophosphate.

inactivation of urease is stronger at pH 9.0 than at 7.6.
Second, this positive influence of the three buffers - although
to a different extent is again stronger with the montmorillonite (Bva-) than with the kaolinite (Bv-) materials. Third,
in the presence of a suitable buffer and pH, urease may even
be reactivated temporarily. This reactivation was much more
pronounced with montmorillonite than with the kaolinite-
containing soil material. Finally, the type of buffer seems
to play an important role. Thus, with the kaolinite-containing
soil, pyrophosphate buffer was most effective, exactly opposite to the montmorillonite Bva-material, which retained the
highest activity in the presence of THAM.

DISCUSSION

Enzymes in soil are exposed to counteracting processes
such as leaching and sorption, hydrolysis and accumulation
or inactivation and stimulation. With respect to the potential activity and ultimate fate of liberated enzymes, their
behavior towards clay minerals is one of the fundamental
aspects, since sorbed enzymes may be protected from
the destructive action of other catalysts by an unknown
shielding mechanism (Burns *et al*., 1972). Such a phenomenon
urges the view of an ever-increasing amount of irreversibly
bound and protected enzymes, particularly in those soils that
are rich in clay minerals (pelozolic soils).
The question of whether sorbed enzymes can be released
or reactivated again upon prolonged incubation with buffer
solutions should be denied now with respect to our studies.
For example, urease, once largely inactivated by attachment
to kaolinite or montmorillonite-containing soil material or
clays, could not be released, neither with water nor by buffer
treatment. Compared to water, however, the inactivation
process of urease was significantly *retarded* in the presence
of a suitable organic or inorganic buffer. This may be
explained by assuming an active competition for attachment
sites between the urease molecule and the buffer compounds.
Nevertheless, no direct relationship was found between the
rate of urease inactivation and the CEC of the clay minerals
used. In terms of soil dynamics, this behavior of urease
could mean that a well-buffered microhabitat is more effective
in preventing enzyme inactivation than a poor, mainly aqueous
solution.
Except for the buffer capacity, the pH of a soil may
affect behavior and activity of urease. Alkaline rather than
neutral conditions significantly delay the inactivation of
urease, although this effect depends on the buffer and type
of clay mineral in question. Thus, at pH 9 the protective
effect of THAM on native urease activity was more pronounced
with montmorillonite soil (Bva) than with the kaolinite

material (Bv). With the Bva material, even a *temporary increase* in urease activity could be measured. At alkaline conditions, montmorillonite is probably more negatively charged than kaolinite; consequently urease was repelled transitorily more intensively by the former than by the latter.

REFERENCES

Burns, R. G., A. H. Pukite andA. D. McLaren. "Concerning the Location and Persistence of Soil Urease," *Proc. Soil Sci. Soc. Am.* 36:308-311 (1972).

Durand, G. "Les Enzymes Dans le Sol," *Rev. Ecol. Biol. Sol.* 2:141-205 (1965).

Kunc, F. and G. Stotzky. "Effect of Clay Minerals on Heterotrophic Microbial Activity in Soil," *Soil Sci.* 118:186-195 (1974).

Martin, J. P., Z. Filip and K. Haider. "Effect of Montmorillonite and Humate Growth and Metabolic Activity of Some Actinomycetes," *Soil Biol. Biochem.* 8:409-413 (1976).

Stotzky, G. "Activity, Ecology, and Population Dynamics of Microorganisms in Soil," *Microbiol.* 2:59-137 (1972).

Skujins, J. "Enzymes in Soils," in *Soil Biochemistry* A. D. McLaren and G. H. Peterson, eds. pp. 371-414.

Tabatabai, M. A. and J. M. Bremner. "Assay of Urease Activity in Soils," *Soil Biol. Biochem.* 4:479-487 (1972).

SECTION V

DESTRUCTION, MINERALYSIS, WEATHERING

CHAPTER 53

SOME ASPECTS OF THE ROLE OF
HETEROTROPHIC MICROORGANISMS IN THE DEGRADATION
OF MINERALS IN WATERLOGGED ACID SOILS

J. BERTHELIN
D. BOYMOND

Centre de Pedologie biologique du C.N.R.S.
B.P. 5 - 54500 Vandoeuvre-les-Nancy, France

In a recent review of Yoshida (1975) concerning different aspects of microbial metabolism of flooded soils, it
appears that if some workers have studied microbial iron and
manganese reduction, little was known concerning the influence
of heterotrophic microorganisms in the weathering processes
of minerals in hydromorphic soils. So microbial processes
of Mn oxidation and reduction (Bromfield and David, 1976)
and of iron reduction (Hamman and Ottow, 1976) by heterotrophic
soil bacteria are now better known than microbial weathering
processes of minerals in the soils.
     In previous experiments (Berthelin and Kogblevi,
1974) we have observed that waterlogging of soil profiles
decreased the aerobic microbial population but promoted the
development of anaerobic bacteria that biosynthesized actively
volatile (formic, acetic and butyric) and semivolatile acids
(lactic). The influence of waterlogging on net microbial
solubilization processes depends upon the chemical nature
of the elements involved. Net microbial solubilization of
Fe and Mn was promoted but was decreased for Si and Al. But,
as in such processes, microorganisms and mineral transformations involved are not known and as solubilization processes
are not well understood, further experiments were conducted
in order to specify some aspects of the role of heterotrophic
microorganisms in the degradation of minerals in waterlogged
acid soils.

MATERIAL AND METHODS

Experimental Device, Soil Composition and Experimental Design

Experiments presented here were based on a method of semicontinuous flow perfusion through soil columns. Experimental devices were similar to those described previously (Berthelin and Kogblevi, 1972, 1974) except for profile models of the third experiment, which were less complex since they were made up of one horizon instead of two. It was shown previously that total sterilization could be achieved by adding (1 g/l final concentration) sodium merthiolate to the nutrient medium percolating through the soil columns.

First experiment: Two factors at two levels in a block of 12 units were studied.

- Waterlogging of soil (a) waterlogged soil columns
                              (b) nonwaterlogged soil columns

- Total sterilization (a) without sodium merthiolate (presence of microorganisms)
                              (b) with sodium merthiolate (1/1000) (absence of microorganisms)

In the first experiment, models of soil profiles were obtained by placing *ca.* 15 g of A horizon, freshly collected from the Hohrodberg brown forest soil (Vosges, France), upon *ca.* 50 g of granite sand from C horizon of the same soil in a glass column 300 mm height and 25 mm in diameter. The samples were sieved so as to eliminate particles coarser than 4 mm in A horizon and 2 mm in C horizon.

Second experiment: Experimental design was the same, except for soil composition. Soil profiles were obtained by placing 45 g of A horizon (freshly collected) upon 150 g of C horizon in a glass column 300 mm height and 45 mm in diameter. Large columns were used in order to place in each column of each treatment (only one column for each treatment) one platimum electrode and one calomel—mercury electrode to measure Eh. The characteristics of A and C soil samples were similar in the first and second experiment. Chemical analyses of horizons are given by Berthelin and Kogblevi (1974).

Third experiment: One factor at three levels (nature of rocks) and one factor at two levels (total sterilization) were studied in a block of 18 waterlogged columns.

            Nature of rocks   (a) granite sand of Hohrodberg (G.S.H.)
                                      (b) granite of Hohrodberg (G.H.)
                                      (c) granite of Ecarpiere (G.E.)

Total sterilization   (a) without sodium merthiolate
(presence of microorganisms)
           (b) with sodium merthiolate
(1/1000) (absence of microorganisms)

In this experiment, conducted to observe the microbial transformations of minerals in waterlogged conditions, model profiles were made up of 50 g each of three types of rock
fragments placed in separate glass columns of 300 mm height
and 25 mm diameter. The rock fragments were particles of
300 to 1000 microns obtained by grinding and sieving two
granites (granite of Hohrodberg, Vosges, France and granite
of Ecarpiere, Maine et Loire, France), or by sieving C horizon samples from brown forest soil of Hohrodberg. The pertinent chemical analyses are given in Table I. All the columns
were waterlogged.

Table I
Chemical Analysis (Percent of Oven Dry Weight)
of Three Types of Rocks (Third Experiment):
Granite of Ecarpiere (G.E.), Granite of Hohrodberg (G.H.),
Granite Sand of Hohrodberg (G.S.H.)

	SiO_2	Al_2O_3	Fe_2O_3	MgO	CaO	MnO	K_2O	Na_2O	TiO_2	Ign. loss
G.E.	71.50	15.50	2.15	0.53	0.58	0.02	3.25	3.95	0.30	2.09
G.H.	69.50	13.50	2.90	2.20	1.40	0.06	2.70	5.80	0.80	1.04
G.S.H.	65.50	14.00	4.50	2.95	1.20	0.08	2.35	4.05	1.00	3.88

Perfusion conditions and nutrient media (M3) (ammonium
phosphate monobasic 0.5 g, asparagine 0.2 g, malt extract
0.2 g, glucose 20.0 g, all made up to 1000 ml with distilled
water) were similar to those described by Berthelin and
Kogblevi (1974). The characteristics of the soil, granite
and granite sand of Hohrodberg were described by Souchier
(1971) and granite of Ecarpiere by Renard (1974). The columns
of rock fragments alone were inoculated with the microflora
obtained from the A horizon of nonsterile waterlogged columns
of the first experiment.

Analytical Methods

Details of the analytical methods for effluents of the
soil columns (mineral elements and organic compounds in
solution) were described by Berthelin *et al.*, (1974),
Berthelin and Dommergues (1976) except for (1) stability of
organomineral complexes that was determined by cation exchange
with resins (Dowex 50 x 8 or Amberlite IR 120) and by stability
of mineral elements in solution at different pH and (2) ferrous

iron (method described by Charlot, 1961). Identification
of microorganisms involved was according to Buttiaux *et al.,*
(1974).

Transformations of minerals were determined (1) by
chemical analysis of rock fragments (Jeanroy, 1972), (2) by
successive chemical extractions with different reagents and
evaluation of weathering index (Berthelin *et al.*, 1974),
(3) by application of X-ray diffraction techniques (Lucas,
1962; Robert and Tessier, 1974) to fine fraction (0-2 μ),
obtained after dispersion and extraction (Rouiller *et al.,*
1972), and (4) by observation of rock fragments with stereoscopic microscope.

RESULTS

Microorganisms Involved and Processes of Solubilization

Waterlogging of soil profiles promoted the development
of anaerobic and facultatively anaerobic bacteria such as
Clostridia [undetermined, but certainly similar to those
characterized by Hamman and Ottow (1976)] and *Bacillus*
(*B. cereus, B. licheniformis, B. polymixa*) that biosynthesized
large amounts of volatile and semivolatile acids (formic,
acetic, butyric, lactic).

After a ten-months perfusion (long-time perfusion),
we observed that such microflora promoted the microbial
solubilization of Fe and Mn and decreased solubilization of
Al (Figure 1). Microbial solubilization of Si, Mg and Ca
was only affected at the beginning of the perfusion (microbial
solubilization of Ca and Mg was promoted and microbial solubilization of Si was decreased).

In the effluents of nonsterile waterlogged columns, all
the minerals elements in solution were exchangeable with
strong cationic resins H^+; only 0 to 10 percent of solubilized
iron was not exchangeable and was under the form of very
stable organometallic complexes. Measurements of ferrous
iron (Fe^{++}) and of the stability of Fe in solution at alkaline
pH showed that 60 to 100 percent of iron was in solution as
Fe^{++}.

In the largest soil columns, the simultaneous measurements of iron in solution in the effluents, of pH of the
effluents and Eh in the columns showed that in nonsterile
waterlogged columns pH decreased between 3.0 to 4.0 and was
not very different from nonsterile nonwaterlogged columns
(Figure 2). In sterile columns, pH was maintained between
4.0 to 5.0; Eh decreased rapidly and was less than -150 mV
after two days perfusion in nonsterile waterlogged columns.
After two months, Eh increased and was maintained near +250 mV
(Figure 2). In nonsterile nonwaterlogged columns, Eh value
decreased not regularly between +425 and +75 mV. In sterile

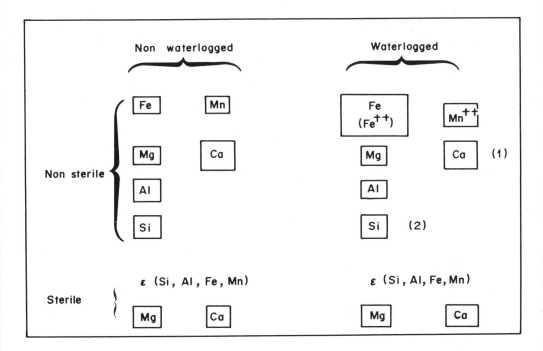

Figure 1. Influence of waterlogging on net microbial solubilization of Si, Al, Fe, Mn, Mg and Ca. The surface of each
rectangle is proportional to the corresponding amount of
solubilized element (ε = very low solubilization).
(1) Waterlogging promoted microbial solubilization of Mg
and Ca at the beginning of perfusion; (2) Waterlogging
decreased microbial solubilization of Si at the beginning
of perfusion.

columns, Eh decreased near +250mV in nonwaterlogged columns,
but near -100 mV in waterlogged columns,
    Solubilization of Fe was very important in nonsterile
waterlogged columns. In one nonsterile waterlogged column,
520 mg of Fe_2O_3 were solubilized, but only 34 mg of Fe_2O_3 in
one nonsterile nonwaterlogged column and 6 to 7 mg of Fe_2O_3
in sterile waterlogged or nonwaterlogged columns. So, despite
the low value of Eh and pH in sterile waterlogged columns,
solubilization of iron was not very important. This result
was important to show the occurrence of heterotrophic
microorganisms (bacteria) as agents of solubilization of iron
in hydromorphic soils.

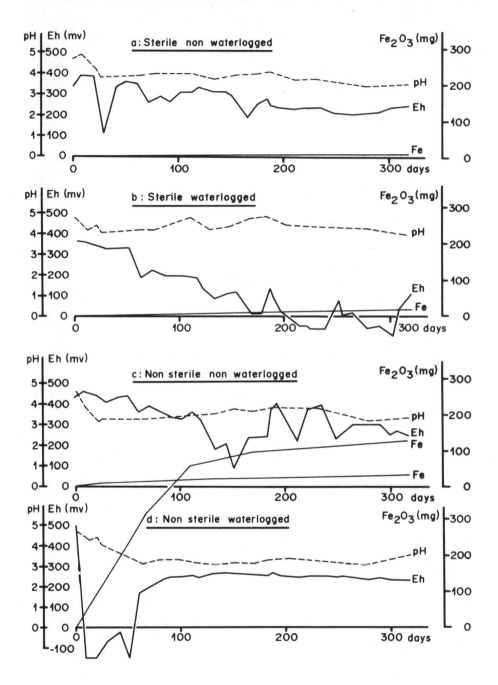

Figure 2.   Influence of waterlogging on soil Eh (mV), on pH
   of effluents and on net cumulative solubilization of Fe
   (mg Fe$_2$O$_3$).   Each column contained 33 g (dry weight) of A
   horizon and 150 g (dry weight) of C horizon.

Microbial Transformations of Phyllosilicates of Three Granitic
Rocks by Anaerobic Microflora in Waterlogged Columns (Third
Experiment)

After a perfusion of 22 weeks, the weights of rock
fragments decreased from 3.6 to 1.8 percent in nonsterile
columns and from 0.2 to 1.2 percent in sterile columns.
Therefore microflora solubilized about 3.4 to 0.6 percent of
minerals. Analysis of mineral elements in solution in the
effluents of the columns (Table II) showed that large amounts

Table II
Cumulative Net Solubilization of Si, Al, Fe, Mg and K
Within 22 Weeks of Perfusion[a]

		SiO_2	Al_2O_3	Fe_2O_3	MgO	K_2O
G.E.[b]	Sterile	70	12	11	19	11
	Nonsterile	242	171	309	79	11
G.H.[c]	Sterile	51	12	12	20	8
	Nonsterile	103	46	67	83	9
G.S.H.[d]	Sterile	40	33	9	30	8
	Nonsterile	170	69	200	86	11

[a]Results expressed as mg SiO_2, Al_2O_3, Fe_2O_3, MgO, K_2O/column.
[b]G.E. = granite of Ecarpiere.
[c]G.H. = granite of Hohrodberg.
[d]G.S.H. = granite sand of Hohrodberg.

of Fe were solubilized by microflora. Microbial solubilization
of Si, Al and Mg were less important. Chemical analysis of
rock or rock fragments after perfusion (results not presented
here) confirmed such microbial solubilization and showed
decreases in the content in Al, Fe, Mn, Mg.
    Microbial solubilization of minerals promoted weather-
ing indices of Si, Al and Fe. Such index was described by
Berthelin *et al.*, (1974). For instance, for granite of
Ecarpiere, granite of Hohrodberg and granite sand of Hohrodberg,
weathering indices of Fe were 4.4, 1.7, 4.8 respectively before
perfusion, 7.2, 2.4, 4.2 respectively after perfusion in
sterile controls, and 31.4, 5.8, 13.4 respectively after perfusion in nonsterile columns.

Microbial Destruction of Primary Chlorite

    In the fine fraction (0-2 µ), X-ray diffraction showed
that chlorite occurred in granite of Ecarpiere (G.E.) before
perfusion and in sterile controls (Figure 3). Chlorite is
characterized by peaks at 7 and 14 Å (peak at 7 Å disappears

Before perfusion

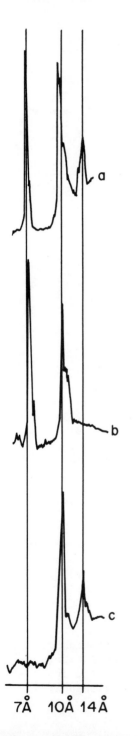

7Å   10Å 14Å

After perfusion

Sterile   Non sterile

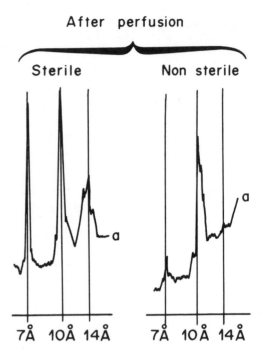

7Å 10Å 14Å   7Å 10Å 14Å

Figure 3. Microbial destruction of chlorite in the fine fraction (0-2 μ) of waterlogged nonsterile columns of granite of Ecarpiere during a 22-week perfusion. X-Ray diffraction diagrams of powder: (a) normal, (b) hydrazine treatment, (c) heating 550° C.

by hydrazine treatment). But chlorite was not present on
X-ray diagram of fine fraction of nonsterile columns (Figure
3). Such results were correlated by decrease of the content
in Fe and Mg and by increase of the content in Si in the
residual fine fraction (0-2 µ) of nonsterile columns (Table
III). Similar modifications of the content in Si (quartz
increases) and in Al and Fe, which decrease, were observed
in the field by Gilkes and Little (1972) and by Coffman and
Fanning (1975) during degradation of chlorite in brown and
leached soils.

Table III
Chemical Analysis (percent of Oven Dry Weight)
of Fine Fraction (0-2 µ) of Granite of Ecarpiere (G.E.),
Granite of Hohrodberg (G.H.)
and Granite Sand of Hohrodberg (G.S.H.) Before and After
a 22-Week Perfusion in Nonsterile Waterlogged Columns[a]

	G.E.		G.H.		G.S.H.	
	Before Perfusion	After Perfusion	Before Perfusion	After Perfusion	Before Perfusion	After Perfusion
SiO_2	51.64	56.64	49.59	49.81	45.05	47.90
Al_2O_3	28.75	27.99	17.30	17.43	24.74	26.31
Fe_2O_3	7.74	3.52	13.38	13.20	15.73	11.47
MnO	0.06	0.02	0.18	0.11	0.43	0.09
CaO	0.32	0.41	10.26	10.33	0.20	0.31
MgO	2.34	1.61	1.15	0.92	7.36	6.74
Na_2O	1.40	1.75	1.45	1.17	0.55	0.74
K_2O	7.39	7.55	4.61	4.85	3.80	3.91
TiO_2	0.35	0.52	2.07	2.24	2.15	2.42
Total	99.99	100.08	99.99	100.00	100.01	99.89

[a]Chemical analysis after a 22-week perfusion in sterile columns was
similar to chemical analysis before perfusion.

Microbial Destruction of Vermiculite

In the fine fraction (0-2 µ) of nonsterile columns of
granite sand of Hohrodberg (G.S.H.), X-ray diffraction showed
a decrease of the peak at 14 Å (Figure 4). This peak is
characteristic of vermiculite because it disappears by heating
at 550° C and does not swell by ethylene glycol treatment.
Such microbial destruction of vermiculite was correlated by
microbial solubilization of Si, Al, Fe and Mg in effluents
and by a decrease of Fe and Mg content of the fine fraction
of nonsterile waterlogged columns of granite sand (Table III).

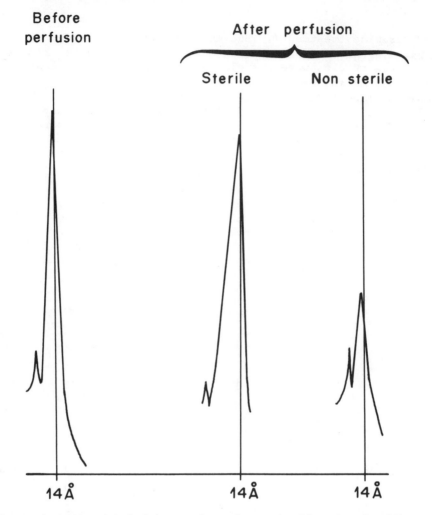

Figure 4.   Microbial destruction of vermiculite in the fine
    fraction (0-2 µ) of granite sand of Hohrodberg (G.S.H.) in
    nonsterile waterlogged columns during a 22-week perfusion.
    X-Ray diagrams of oriented aggregates Mg saturated.

Transformation of Interstratified Minerals (Illite-Vermiculite)
into Montmorillonite

        In the fine fraction(0-2 µ) of granite of Hohrodberg,
X-ray diffraction showed the presence of interstratified
minerals (illite-vermiculite) before perfusion and in sterile
controls after perfusion (not presented here). After a 22-
week perfusion, we observed that a peak at 14 Å increased in
nonsterile columns. Before perfusion such a peak does not
swell by ethylene glycol and partially disappears by heating
at 550° C (a part of the peak at 14 Å contains chlorite and
the absence of chlorite weathering could be dependent on size

of clay particles). But, after perfusion in nonsterile waterlogged columns, a part of the important 14 Å peak swells by
ethylene glycol treatment and is characteristic of montmorillonite (Figure 5). So, in our experimental conditions,
microbial activity in waterlogged columns transformed (illite-
vermiculite) into "montmorillonite of degradation."

Transformation of Biotite of Granite of Hohrodberg

Stereoscopic microscope observations of biotite grains
showed that in nonsterile waterlogged columns, some biotite
grains became partially and/or entirely white and brittle.
As during the processes of weathering induced by partial
sterilization of soils (Berthelin and Belgy, in press) such
colorless biotite grains could be Si-Al residue. Evaluation
of total solubilization of ferromagnesian minerals (here only
biotite) showed that 7.0 to 8.0 percent of biotite was entirely
solubilized in nonsterile columns, but only 1.5 to 2.0 percent
in sterile columns.

CONCLUSION

Waterlogging of soil profiles containing granitic rocks
promoted the development of an anaerobic microflora. Microorganisms involved were *Clostridia* and *Bacillus* (*B. cereus,
B. polymyxa, B. licheniformis*). This anaerobic and facultatively anaerobic microflora solubilized mineral elements by
reduction of Fe and Mn and by "acidolysis," that is, by
production of volatile acids (formic, acetic, butyric) and
semivolatile acids (lactic) forming salts or complexes salts
with all mineral elements (mineral elements in solution were
extracted by strong cationic resin). Microbial activity
enhanced significantly gleyification and solubilization of Fe
and Mn under the reduced ionic state. Microbial solubilization
of Ca and Mg was not significantly affected by waterlogging
except at the beginning of perfusion, but Si and Al microbial
solubilization was decreased, most particularly at the beginning
of the perfusion. These elements were under ionic form.
During microbial solubilization of mineral elements, primary
and secondary phyllosilicates of three granitic rocks were
transformed.

Primary chlorite of granite of Ecarpiere (granite with
two micas with biotite chloritization and plagioclases hema-
tization) was continuously and entirely solubilized. Such
destruction of primary chlorite was observed in different
soils during weathering of loess in lehm (Millot and Camez,
1956) or during podzolization processes (Arnaud and Mortland,
1963; Coen, 1970; Seddoh and Pedro, 1974). Schwertmann (1976)
specified that in the field, chlorite was transformed into
vermiculite in a slightly acid environment poor in organic

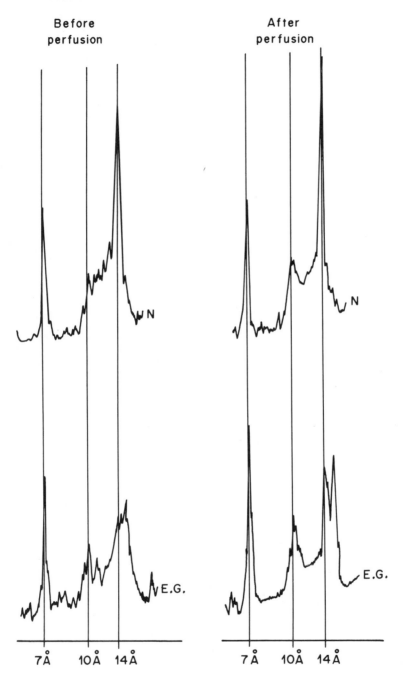

Figure 5. Microbial transformation of (illite-vermiculite)
into montmorillonite of degradation in the fine fraction
(0-2 μ) of granite of Hohrodberg (G.H.) in nonsterile waterlogged columns. X-Ray diagrams of powder: N normal, E.G.
ethylene glycol. (X-Ray)diagrams of sterile columns after
perfusion were similar to those before perfusion.)

matter but was entirely destroyed in an acidic environment rich in organic matter.

In the granite sand of Hohrodberg, vermiculite was continuously solubilized and in the granite of Hohrodberg (illite-vermiculite) was transformed into montmorillonite of degradation. Such results are to be related to Gury's (1976) observations, which showed that during hydromorphic degradations of soils by acid organic water table (illite-vermiculite) were transformed into montmorillonite by degradation as defined by Schwertmann (1962), Gjems (1962) and Coen (1970) and that vermiculite disappeared. The biotite of Hohrodberg granite was solubilized by microbial activity in waterlogged acid soils or could be transformed into a white brittle residue.

Such processes should be important in anaerobic horizons of degraded paddy soils, in reduced niches containing organic matter and could participate to the degradation of hydromorphic soils by acidic organic reducing water table. Such degradations have been described by Bernot (1965) in the presence of a perched water table, Juste (1965) and Gury (1976) in the presence of an acid organic water table along terraces. Since Fe reduction did not occur in sterile controls, even when pH and Eh were low, heterotrophic microorganisms were thought to play a major role in the reduction of Fe in waterlogged conditions.

REFERENCES

Arnaud, R. J. St. and M. M. Mortland. "Characteristics of the Clay Fractions in a Chernosemic to Podsolic Sequence of Soil Profiles in Saskatchewan," *Canad. J. Soil. Sci.* 43:336-349 (1963).

Bernot, J. "Contribution a l'etude des sols lessives des terrasses de la Garonne," *Ann. E.N.S.A. Toulouse* 10:1-101 (1965).

Berthelin, J. and A. Kogblevi. "Influence de la sterilisation partielle sur la solubilisation microbienne des mineraux dans les sols," *Rev. Ecol. Biol. Sol* 9(3):407-419 (1972).

Berthelin, J. and A. Kogblevi. "Influence de l'engorgement sur l'alteration microbienne des mineraux dans les sols," *Rev. Ecol. Biol. Sol* 11(4):499-509 (1974).

Berthelin, J., A. Kogblevi and Y. Dommergues. "Microbial Weathering of a Brown Forest Soil - Influence of Partial Sterilization," *Soil Biol. Biochem.* 6(6):393-399 (1974).

Berthelin, J. and Y. Dommergues. "The Role of Heterotrophic Microorganisms in the Deposition of Iron and Carbon in Soil Profiles

Berthelin, J. and Y. Dommergues. "The Role of Heterotrophic Microorganisms in the Deposition of Iron and Carbon in Soil Profiles," in *Environmental Biogeochemistry*, Vol. 2, J. O. Nriagu, ed. (Ann Arbor, Michigan: Ann Arbor Science Pub., 1976), pp. 609-631.

Berthelin, J. and G. Belgy. *Microbial Degradation of Phyllosilicates During Podzolization Processes Induced by Partial Sterilization* (in press).

Bromfield, S. M. and D. J. David. "Sorption and Oxidation of Manganous Ions and Reduction of Manganese Oxide by Cell Suspensions of Manganese Oxidizing Bacterium," *Soil Biol. Biochem.* 8:37-43 (1976).

Buttiaux, R., H. Beerens and A. Taxquet. *Manuel de Techniques Bacteriologiques* (Paris: Flammarion Medecine Sciences, 1974).

Charlot, G. *Les methodes de la chimie analytique. Analyse quantitative minerale* (Paris: Masson and Sie, 1961).

Coen, G. M. "Clay Mineral Genesis in Some New York Spodosols," *Diss. Abstr.* 31B:17-18 (1970).

Coffman, C. B. and D. S. Fanning. "Maryland Soils Developed in Residuum from Chloritic Metabasalt Having High Amounts of Vermiculite in Sand and Silt Fractions," *Soil Sci. Soc. Amer. Proc.* 39(4):723-732 (1975).

Gilkes, R. J. and I. P. Little. "Weathering of Chlorite and Some Associations of Trace Elements in Permian Phyllites in Southeast Queensland," *Geoderma* 7(3-4):233-247 (1972).

Gjems, O. "A Swelling Clay from the Weathering Horizon of Podsol," *Clay Miner. Bull.* 5:28 (1962).

Gury, M. "Evolution des sols en milieu acide et hydromorphe sur terrasses alluviales de la Meurthe," *These Doct. Spec. Pedol.*, Univ. Nancy I. (1976).

Hamman, R. and J. C. G. Ottow. "Isolation and Characterization of Iron Reducing Nitrogen-Fixing Saccharolytic *Clostridia* from Gley Soils," *Soil Biol. Biochem.* 8:357-364 (1976).

Jeanroy, E. "Analyse totale des silicates naturels par spectrophotometrie d'absorption atomique. Application au sol et a ses constituants," *Chim. Anal.* 54(3):159-166 (1972).

Juste, C. "Contribution a l'etude de la dynamique de l'aluminium dans les sols acides du Sud-Ouest atlantique: application a leur mise en valeur," *These Doct. Ing.* Fac. Sci. Nancy (1965).

Lucas, J. "La transformation des minéraux argileux dans la sédimentation. Etudes sur les argiles du Trias," *Mem. n° 23, Bull. Serv. Carte Geol. Als. Lorr* (1962).

Millot, G. and Th. Camez. "Evolution des minéraux argileux dans les loess et les lehms d'Achenheim (Alsace)," *C.R. VIth Intern. Congr. Soil Sci.,* Paris (1956).

Renard, J. P. "Etude pétrographique et géochimique des granites du district uranifère de Vendée: liaisons entre l'évolution mineralogique et le comportement de l'uranium; conséquences pour la prospection," *Doc. C.E.A.-R-4407 C.E.N. Saclay* (1974).

Robert, M. and D. Tessier. "Méthode de préparation des argiles des sols pour des études minéralogiques," *Ann. Agron.* 25(6):859-882 (1974).

Rouiller, J., G. Burtin and B. Souchier. "La dispersion des sols dans l'analyse granulométrique. Méthode utilisant les résines échangeuses d'ions," *Bull. E.N.S.A. Nancy* 14(2):193-205 (1972).

Schwertmann, U. "Eigenschaften und Bildung auf weitbarer (Quellbarer) Dreischicht. Tonminerale in Böden aus Sedimenten," *Beitrage Z. mineral. Petrogr.* 8:1-9-209 (1962).

Schwertmann, U. "Die Verwitterung mafischer Chlorite," *Z. Pflanzenern. Bodenk.* 1(1):27-36 (1976).

Seddoh, F. and G. Pedro. "Caractérisation des différents stades de transformation des biotites et biotites chloritisées dans les arènes granitiques du Morvan," *Bull. Groupe Fr. Argiles* 26(1):107-125 (1974).

Souchier, B. "Evolution des sols sur roches cristallines à l'étage montagnard (Vosges)," *Thèse Doct. Etat,* Univ. Nancy I - *Mem. n° 33, Bull. Serv. Carte Geol. Als. Lorr.* (1971).

Yoshida, T. "Microbial Metabolism of Flooded Soils," in *Soil Biochemistry* Bol. 3, E. A. Paul and A. D. McLaren eds., (New York: Marcel Dekker, Inc., 1975), pp. 83-122.

CHAPTER 54

MICROORGANISMS AND WEATHERING OF A SANDSTONE MONUMENT

F. E. W. ECKHARDT

Institut für Allgemeine Mikrobiologie
Universität Kiel
Olshausenstr. 40-60
D 2300 Kiel, Federal Republic of Germany

INTRODUCTION

Weathering of stone in buildings and monuments will proceed more or less quickly due to such well known environmental agencies as temperature, humidity or salt concentration. In addition, the microbial population of the stone may also contribute to its decay. Summaries on the weathering processes of building material are given by Kaiser (1930), Kieslinger (1932), and de Quervain (1945). They all stress the physical and chemical agents as most important for deterioration.

Microbiological aspects of weathering of building stones are outlined in other reviews (Paine *et al.*, 1933; Krumbein, 1966, 1968, 1973; Pochon *et al.*, 1968; Hueck-van der Plas, 1968; Jaton, 1971). In summary, the production of acids by microorganisms living on the stone seems to be the most important deteriorating agent. Sulfuric or nitric acids are produced by thiobacilli or nitrifying bacteria within weathering stones. However, organic acids such as oxalic, citric or gluconic acids, excreted by heterotrophic bacteria and fungi, are thought to be more important weathering agents. The presence of microorganisms, even in great numbers, on weathered material does not necessarily imply that these organisms have caused the damage observed; however, these populations may be involved in enhancing abiotic processes.

Air pollutants, such as sulfur dioxide, have been found to be the main influence in deterioration of historically important buildings like the cathedral of Cologne (Luckat, 1973-1976). However, severe weathering has also been found

in monasteries, castles or sculptures far away from any urban
or industrial air pollution (Riederer, 1972, 1973). Pochon
and co-workers (1960, 1968) demonstrated the influence of
thiobacilli on weathering the temples of Angkor in Cambodia.
Furthermore, the restoration of the Wolfsburg Castle was only
successful when the walls were treated with an antibacterial
agent that presumably killed nitrifying bacteria within the
walls of the castle (Riederer, 1972).

Since the activities of microorganisms are suspected
to be an important factor in the weathering of building
stones, the present work was undertaken to find out whether
the decay of a 150-year-old sandstone obelisk could be related
to activity of its microbial population.

MATERIAL AND METHODS

The monument under study is about 4 m high and built
from several angular blocks of a grey sandstone, with a ball
at its top. This memorial was erected in 1830 in the center
of an intersection in Kiel. Today, its surface is covered
with a black layer, which is continuous on the northeastern
side of the monument. A crust of about 1-2 mm thickness on
the surface of the stone is scaling off at different places,
especially on the northeastern side.

Background environmental data were gathered by
measuring temperature, light intensity and concentrations of
carbon monoxide and hydrocarbons in the air. The temperature
at the surface of the stone was measured with a single channel
telethermometer (YSI, model 43 TD), equipped with a thermistor
probe "Banjo," type 408. Light intensity at the same locations was measured with a lux meter (Gossen, type Trilux).
The levels of the air pollutants, CO and hydrocarbons in the
rather heavily travelled intersection were determined with a
Multi-Gas Detector, model 21/31 (Drägerwerke, Lübeck).

The pH, water content and percent organic matter of
the crust material were determined at different sites on the
monument. The pH was measured in a suspension of ground
material (1:2.5, w/v; Scheffer-Schachtschabel, 1970) with a
Pye Model 290 pH Meter 24 hr after the material was suspended
in 0.1 N KCl. The water content and amount of organic matter
were determined by weight after drying at 105° C for 24 hr
and after dry combustion at 600° C for 5 hr, respectively.
The mineral composition of thin sections of the stone was
determined by light microscopy and by electron microprobe
analysis (Siemens, type Mikrosonde). Scanning electron microscopy of crust material was done with a Cambridge Stereoscan
Mk II.

Samples of the crust material were examined for total
numbers of heterotrophic bacteria and fungi by plating dilutions on nutrient agar (Difco), nutrient agar plus 1% glucose,

and on malt extract agar (Merck). Also, crust material was ground in a mortar and spread on the same agar media. The presence of sulfur-oxidizing and nitrifying bacteria was determined by inoculating appropriate autotrophic growth media (Krumbein, 1966) with dilutions of crust material.

The distribution of the microorganisms on the surface of the stone was investigated using a new agar impression technique, which was suitable for sampling smooth as well as rough or even angular surfaces. The following procedure was employed. A double layer of surgical gauze was cut into square pieces about 5 cm on a side. The gauze was placed on a square of aluminum foil about 1 cm larger, and both items were sterilized in glass petri dishes. About 5 ml of desired agar medium, sterilized in tubes and cooled to about 45° C was then poured aseptically onto the gauze. For inoculation, the impregnated gauze was removed from its petri dish together with the underlying foil, using sterile forceps. The agar was impressed onto the surface, protected from contamination by the foil. Next, the agar and foil were removed and the inoculated agar, without the foil, was placed in a sterile plastic petri dish and incubated at room temperature. To avoid desiccation of the thin agar layer, the dishes were incubated in a humid chamber. This technique was suitable for revealing the local distribution of microorganisms on the surface; however, it was not suitable for quantitative estimations.

Isolates, obtained from these studies, were examined for acid production by growing them in a medium containing 0.5% glucose in a mineral salts solution, consisting of 0.05% $(NH_4)_2SO_4$; 0.05% $K_2HPO_4 \cdot 3\ H_2O$; and of 0.02% $MgSO_4 \cdot 7\ H_2O$.

RESULTS

The data obtained from measuring the temperature and intensity of light at the surface of the monument showed no great differences, except during exposure to direct sunshine (Tables I and II). Normally, the temperature at the surface of the stone was slightly higher than that of the air.

According to a traffic census in 1973, more than 34,000 car units passed this intersection every 24 hr. But the pollution coming from exhaust gases was quite low in front of the monument. Carbon monoxide was present at a level of 1 ppm/hr, whereas the amount of carbohydrates could not be quantitated at the lowest sensitivity level of 3 mg/l of the instrument used. These surprisingly low levels might have been due to the wind, which is blowing almost constantly at this location.

The pH and water content were different at the various sampling sites, whereas the percent of organic matter was relatively consistent (Table III). The extraordinarily high

Table I
Temperature and Light Intensity at the Surface of the
Monument at Different Levels above Base at the
Northeastern Side

	June 1		June 14	June 22	
	Lux	°C	°C	°C	Lux
Level					
160 cm	8000	16.2	–	–	8000
200 cm	12000	16.5	24.0	–	8000
310 cm	13000	16.5	24.0	21.0	8000
Air[a]	–	15.5	22.5	18.1	–
Hour of day	1:00 p.m.		5:00 p.m.	10:00 a.m.	
Weather	cloudy, wind		cloudy, wind	sunshine, wind	

[a]5 cm from surface.

Table II
Temperature and Light Intensity on Different Sides
of the Monument at the 310-cm Level

	NE	SE	SW	NW
Air[a], °C	18.8	21.0	17.9	17.0
Surface, °C	21.0	31.0/29.0[b]	18.0	17.8
Surface, lux	8000	56 00	8000	7000

[a]5 cm from surface.
[b]Microprobe protected from sunshine.

Table III
Water Content, pH and Percent Organic Matter
Within Crust Material

Sample	Water Content %	pH	Organic Matter %
Obel "2"	4.25	5.4	2.5
Obel "5"	1.54	9.6	2.2
Obel "7"	0.70	4.9	1.7
Obel "9"	0.50	4.6	1.8
Obel "11"	5.3	5.7	–

pH (9.6) at position "5" could have been derived from the alkaline mortar (pH 9.3) of the seam above this sampling site. The higher water contents of samples taken at positions "2" and "11" may have resulted from the absence of a well-developed crust on the surface at these sites.

When thin sections of the sandstone were examined, a dense packing of quartz grains could be seen. These were mixed with a few grains of feldspar, muscovite and some ziosite within a thin layer of a clayey and siliceous cementing material. The mineral composition of the ball at the top of the obelisk differed from the other blocks. It showed a rather thick layer of cement with much hornblende among the quartz grains. Much kaolin was also found within this material.

Electron microprobe analysis revealed that clay was the main amorphous substance between the quartz grains and that considerable amounts of gypsum existed within the crust. The black coating layer of the crust was composed of iron oxide and/or chlorite. Only small quantities of lead were detected.

Electron microscopy confirmed the findings of the dense packing of quartz grains and of the existence of a coating on the outer surface of the stone. In addition, the presence of microorganisms was demonstrated on the inner and outer surfaces of crust material as well as within the crust layer. Mycelia and microorganisms like coccoid algae could be seen (Figures 1 and 2) in addition to coatings and slime tracks (Figure 3), which may have been of bacterial origin. However, bodies of bacteria could not be definitely identified. In several instances, mycelia were seen with mineral coatings (Figure 4). Typical formations of kaolin were also seen beside mycelia and large coccoid organisms (Figure 2), suggesting that transformation of minerals may be affected by these microorganisms.

Concerning the microbial population of the monument, many different bacteria, yeasts and filamentous fungi were cultivated by the techniques employed. The total numbers of heterotrophic bacteria within crust material were similar at the different locations studied, whereas the numbers of yeasts and filamentous fungi, taken as a group, were more dispersed (Table IV). No growth occurred in the dilutions made to count thiobacilli and nitrifying bacteria; however, microorganisms resembling these bacteria were present in enrichment cultures.

When the surfaces were sampled with the agar impression technique, results different from those obtained by cultivating ground material of the crust layer were obtained. Several yeast species grew from both ground crust material and the inner surface of the crust; however, almost none of these yeasts could be cultivated from samples of the outer surface of the crust. On the other hand, a large number of colonies of filamentous fungi grew in samples from the outer surface of the crust, but no colonies were obtained from the inner

Figure 1.   Structures resembling fungal mycelia within cross-
    sectioned crust material.   SEM.   One side of the photograph
    corresponds to 100 μm.

side.   Bacterial colonies grew in appreciable numbers in
samples from both sides of the crust and from ground material.
By comparing the types of colonies that grew from the outer
surface with colonies that grew on agar plates exposed to the
air, airborne contaminants were identified.
        In these studies, 75 bacterial strains, 10 yeast
species and 20 species of filamentous fungi were isolated in
pure culture.   We are currently in the process of characteri-
zing these isolates.   Preliminary attempts were made to
check isolates for their weathering activities on the monument material by testing for acid production.   The isolates
were cultivated in tubes containing 5 ml of 0.5% glucose-
mineral-salts-solution, with and without addition of 100 mg
of ground monument material.   After six to ten weeks of
incubation at room temperature, all of the yeasts and filamentous fungi had lowered the pH of the solution from 5 to
around 3, but only 2 of the bacterial strains lowered the
pH.   The addition of monument powder had no great influence
on acid production.

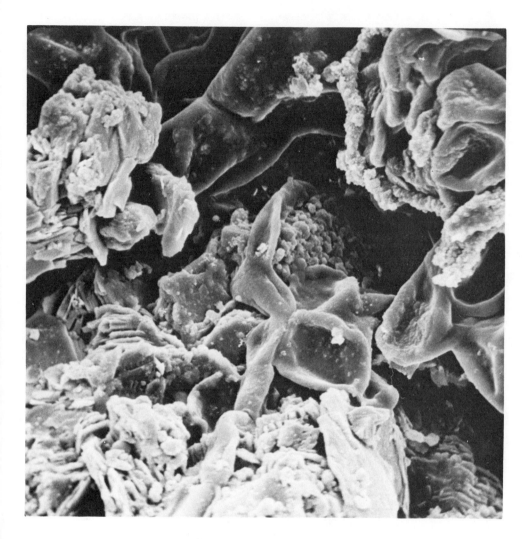

Figure 2. Structures resembling fungal mycelia and coccoid microorganisms on crust material. Kaolin formations (left) and encrusted mycelia (upper right). SEM. One side of the photograph corresponds to 50 μm.

Thin layer chromatography of the medium in which the yeast and fungus strains had grown revealed the presence of citric, oxalic, gluconic and glucuronic acids. In addition, extracts of crust material from several sites of the monument showed the presence of oxalic and citric acids.

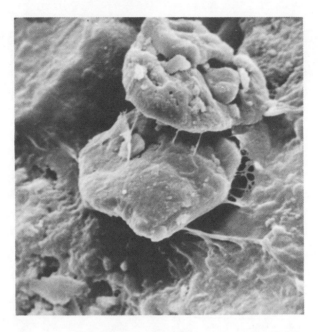

Figure 3.   Slime tracks and coatings on weathered crust material.
SEM.   One side of the photograph corresponds to 20 μm.

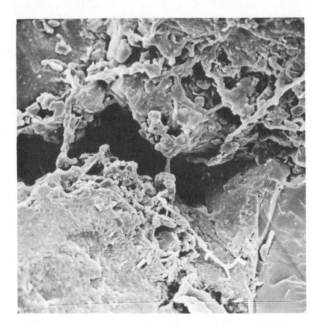

Figure 4.   Mineral coatings on fungal mycelia within the crust.
SEM.   One side of the photograph corresponds to 100 μm.

Table IV
Total Numbers of Heterotrophic Bacteria
and Fungi per gram of Crust Material

	Bacteria	Fungi
Obel "3"	2.9×10^4	4.6×10^4
Obel "9"	2.7×10^3	2.4×10^3
Obel "11"	2.1×10^4	9.6×10^5

DISCUSSION

The studies on the microbial population of the sand-stone monument demonstrated the presence of a rather high number of different microorganisms. Most of the bacteria that grew in agar impression cultures seemed to be contaminants from the air. This was in contrast to yeasts and fungi, most of which seemed to inhabit the stone proper. These results are in agreement with those of Barcellona *et al.*, (1972), Krumbein (1973) and Pochon *et al.*, (1974), who stated that fungi are the predominant microorganisms found on stones of buildings and monuments. The presence of oxalic and citric acids within samples of the monument as well as in the growth medium inoculated with isolated yeast and fungal species suggests that such heterotrophs excreted these organic acids while growing on the monument. Organic acids can contribute to weathering by acidifying the environment and by chelating cations (Henderson *et al.*, 1963; Muller *et al.*, 1963, 1964; Schalscha *et al.*, 1967). Keller (1956) as well as Oberlies *et al.*, (1964) showed that microorganisms are active in transforming feldspar and other aluminum silicates to kaolin. Our electron micrographs show the presence of kaolin near mycelia (Figure 2). Mycelia (Figure 4) that are encrusted with minerals, indicating the transformation of components of the monument, may also have contributed to the weathering of this monument.

In summary, this work has shown that weathering of this sandstone monument was probably affected by microorganisms, which enhanced the deterioration of both the cementing material and the aluminum silicates by producing organic acids.

ACKNOWLEDGMENTS

I am grateful to H. D. Pokojski for his valuable technical assistance and Elke Heiden for her assistance with photographic work. I wish to acknowledge D. Ackermand for help with electron microprobe analysis and C. Samtleben and W. Reimann for help with scanning electron microscopy. I am indebted to W. C. Ghiorse for helpful discussions of the manuscript.

REFERENCES

Barcellona, S., L. Barcellona-Vero and F. Guidobaldi. *The Front of S. Giacomo Degli Incurabili Church in Rome: Chemical and Biological Surface Analysis*, Publication Nr.15, (Rome: Istituto di fisica tecnica- Consiglio nazionale delle Ricerche, 1972).

Henderson, M. E. K. and R. B. Duff. "The Release of Metallic and Silicate Ions from Minerals, Rocks, and Soils by Fungal Activity," *J. Soil Sci.* 14:236-246 (1963).

Hueck-van der Plas, E. H. "The Microbiological Deterioration of Porous Building Materials," *Int. Biodetn. Bull.* 4:11-28 (1968).

Jaton, C. *Contribution a l'étude de l'altération microbiologique des pierres de monuments en France,* Thesis, Paris (1971).

Kaiser, E. "Über eine Grundfrage der natürlichen Verwitterung und die chemische Verwitterung der Bausteine im Vergleich mit der in der freien Natur," *Chemie der Erde* 4:290-342 (1930).

Keller, W. D. "Clay Minerals as Influenced by Environments of their Formation," *Bull. Amer. Assoc. Petrol. Geol.* 40:2689-2710 (1956).

Kieslinger, A. *Zerstörungen an Steinbauten. Ihre Ursachen und ihre Abwehr.* (Leipzig u. Wien: F. Deutick, 1932).

Krumbein, W. E. *Zur Frage der Gesteinsverwitterung (Über geochemische und mikrobiologische Bereiche der exogenen Dynamik).* Dissertation, Wurzburg (1966).

Krumbein, W. E. "Zur Frage der biologischen Verwitterung: Einfluß der Mikroflora auf die Bausteinverwitterung und ihre Abhängigkeit von edaphischen Faktoren," *Z. allg. Mikrobiol.* 8:107-117 (1968).

Krumbein, W. E. "Uber den Einfluß von Mikroorganismen auf die Bausteinverwitterung - eine ökologische Studie," *Dtsch. Kunst- und Denkmalpflege* 31:54-71 (1973).

Luckat, S. "Luftverunreinigungen als Ursache des Steinzerfalls?" *Dtsch. Kunst- u. Denkmalpflege* 31:45-50 (1973a).

Luckat, S. "Die Einwirkung von Luftverunreinigungen auf die Bausubstanz des Kölner Domes, I - III," *Kolner Domblatt* 36/37:65-74 (1973b); 38/39:95-106 (1974); 40:75-108 (1975).

Luckat, S. "Stone Deterioration at the Cologne Cathedral Due to Air Pollution," in *The Conservation of Stone*, R. Rossi-Manaresi, ed. (Bologna: Centro per la Conservazione della Sculture All'Aperto, 1976), pp. 37-53.

Muller, G. and I. Förster. "Der Einfluß mikroskopischer-Bodenpilze auf die Nahrstofffreisetzung aus primären Mineralien, als Beitrag zur biologischen Verwitterung," *Zbl. Bakt.* II 116:372-409 (1963); 118:594-621 (1964).

Oberlies, F. and J. Zlatanovic. "Uber ein illitisches Tonvorkommen in Jugoslawien und Betrachtungen zu seiner Entstehung," *Ber. Dtsch. Keram. Ges.* 41:691-695 (1964).

Paine, S. G., F. V. Linggood, F. Schimmer and T. C. Thrupp. "The Relationship of Microorganisms to the Decay of Stone," *Phil. Trans. Roy. Soc.*, Ser. B 222:97-127 (1933).

Pochon, J., P. Tardieux, J. Lajudie, M. Charpentier, J. Delvert, R. Trian and M. Bredillet. "Degradation des temples d'Angkor et processus biologiques," *Ann. Inst. Pasteur* 98:457-461 (1960).

Pochon, J. and C. Jaton. "Facteurs biologiques de l'Alteration des Pierres," in *Biodeterioration of Materials*, A. H. Walters and J. E. Elphick, eds. (Amsterdam: Elsevier Publ., 1960) pp. 258-268.

Pochon, J., L. Barcellona, C. Giacobini, M. A. Chalvignac, C. Jaton and G. Torraca. *Etude des facteurs biologiques d'altération de la Pietra D'Istria a Venise.* Report to UNESCO, unpublished. (Rome: Internat. Centre for Conservation, 1974).

Quervain, F. de. "Verhalten der Bausteine gegen Verwitterungseinflüsse in der Schweiz. Teil I. Beitr. z. Geol. d. Schwiez," *Geotechn.* Ser. 23:1-56 (1945).

Riederer, J. "The Conservation of German Stone Monuments. in *The Treatment of Stone*, R. Rossi-Manaresi and G. Torraca, eds. (Bologna: Centro per la Conservazione delle Sculture all'Aperto, 1972) pp. 105-138.

Riederer, J. "Die Wirkungslosigkeit von Luftverunreinigungen beim Steinzerfall," *Staub-Reinhaltung d. Luft* 33:15-19 (1973).

Schalscha, E. B., H. Appelt and A. Schatz. "Chelation as a Weathering Mechanism," *Geochim. Cosmochim. Acta* 31:587-596 (1967).

Scheffer-Schachtschabel. *Lehrbuch der Bodenkunde* (Stuttgart: Enke - Verlag, 1970).

CHAPTER 55

DECAY OF PLASTER, PAINTINGS AND WALL MATERIAL
OF THE INTERIOR OF BUILDINGS VIA MICROBIAL ACTIVITY

W. E. KRUMBEIN
C. LANGE

Universität Oldenburg
Postfach 25 03
D-2900 Oldenburg, Federal Republic of Germany

INTRODUCTION

Early in the period of the Industrial Revolution and
technical civilization, in terms of the high energy consuming
western European and North American technology, it was stated
that in contrast to any other technology prior to the present
one the increased rate of decay of cultural treasures is a
still growing danger (Julien, 1882). Kieslinger (1932),
Winkler (1975), and Weber (1977) have stressed that decay of
building stones is a natural phenomenon which has always
endangered human stone constructions, objects of art and
mineral paints. This can be shown by early reports (Moses
III, Herodot II).
    On the other hand there is no doubt that the industrial
revolution with its deep influence on biogeochemical cycles
and geomicrobiological processes has accelerated stone decay
and added new dimensions to the problem. Pochon (1948),
Kauffman (1960), Krumbein (1966, 1968, 1972), and others have
stressed the mutual relation between man's industrial effects
and increased weathering by emmissions of CO_2, nitrogen and
sulfur compounds on rock outside and inside of buildings.
Only in the past ten years has man become aware of these
effects, and scientific studies were started. Some drastic
effects resulted — for example, the closing of the cave of
Lascaux because human access was the cause of heavy decay
processes on the prehistoric paintings of the Magdalenien.
    Later, UNESCO programs (Angkor Vat, Borobodur, Venice)
and the efforts of new emerging industrial branches have
shown that more elaborate study of the related decay processes

as well as the development of protecting and conserving
methods are necessary in future.   Mechanical, chemical, biogeochemical and geomicrobiological processes are increased
by many of the environmental pollutants.   The two combine to
form extremely fast decay rates, mainly in the vicinity of
industrial areas and human agglomerations.

The methodology of weathering research as well as
protecting methods must develop to a much higher extent if
the treasures of art and architecture man has produced during
the past millenia are to be preserved.   Literature surveys
and reviews have been made by Krumbein (1972), Jaton (1971),
Berthelin (1976) and Riederer (1973).

METHODOLOGY

Several different methods and methodological approaches
have been used to investigate building stone deterioration.

1) Physical and physical-chemical methods in the
   immediate environment of the building or monument
   in question.   These have been described extensively
   by Riederer (1973) and Kessler (1960).   The measurement of temperature, humidity, light, rain and wind
   exposition, as well as data on exposure to emission
   sources and emission measurements, are important
   (Luckat, 1973).

2) Laboratory measurements such as porosity, durability
   tests and standard tests of the National Bureau of
   Standards or according to any other industrial norm
   tests are as important as the measurement of pH,
   electrolytes and fractionate dissolution of cations
   and anions with cold and warm water, diluted and
   concentrated acid (Kessler, 1960; Krumbein, 1966;
   Marschner, 1973; Luckat, 1972, 1973).

3) Classical mineralogical and petrological methods
   such as thin sectioning and analyses, X-ray analyses
   of weathering crusts and natural rock, and total
   geochemical analysis of fresh rock in comparison to
   altered material have to be considered as well.
   These methods are general methods of mineralogy
   and geochemistry and are described in many textbooks.

4) Microbiological methods.   The microbiology of weathering rocks can be studied by methods derived from soil
   microbiology or general microbial ecology.   They
   include: enrichment techniques (Aaronson, 1970),
   counting and plating techniques (Krumbein, 1966),
   quantitative percolation tests, and batch culture
   tests (Berthelin, 1976), as well as many modern
   methods in microbial ecology such as absorption tests,
   microscopy, scanning electron microscopy, fluorescent
   microscopy, autoradiography, X-ray analyses and
   continuous culture (Rosswall, 1972).

In this chapter we survey the weathering capability of a
microflora developed on the inside walls of a church in
northwest Germany. The walls were exposed to dim light and
relatively high humidity by water penetrating through the
walls and from the floor.

RESULTS

The church walls were examined macroscopically
(Figures 1, 2 and 3). It was evident from this survey that
humidity and light were the controlling factors of the growth
of microorganisms on the walls of the church. Sections of
the plaster cover, paintings and the rock crevices were taken
to the laboratory and analyzed by SEM methods, electron dis-
persive X-ray analysis and culture methods.
The microscopic analysis of the patches showed clearly
that green algae, coccoid and filamentous blue-green algae
(*cyanobacteria*) and some diatoms were closely associated with
a number of fungi. Lichenization of the fungal and algal
flora was detected macroscopically. Culture assays imitating
the natural conditions of the inside walls showed that the
algal colonies were always associated with fungi. On agar
dishes vertical growth of algae was initiated by the fungal
mycelia. Changes in humidity of the incubator atmosphere
as well as changes in light and temperature were not analyzed
quantitatively. From initial orientation experiments it could
be shown that algal and fungal growth were ruled by these
factors and that periods of lower humidity favored decay of
many of the organisms. In exchange, this was producing
an increase in slime formation and the bacterially associated
flora.
Increased light caused penetration into the media and
increased growth into the rock crevices and pores. Decreasing
light initiated massive growth of cyanophytes. Several of
the cyanophytes seem to have boring capabilities of mechanical
properties since rock particles and plaster were penetrated.
Periods of desiccation alternating with wet periods
enhanced slime production and the development of mucous
coatings on the minerals (Figures 4 and 5). EDX analyses of
several parts of the wall cover with the associated flora
indicated material transfer from rocks to algal, bacterial
and fungal flora. These analyses were not conducted to
quantitative results so far. It was shown, however, that
algae and fungi in contact with different particles expose
different levels of phosphorus, calcium, natrium, kalium, iron,
and several trace elements. In Figures 6 and 7 simple semi-
quantitative linescans of algae associated with weathering
rock particles are shown to demonstrate the advantages of
the method of combining SEM with analytical methods.

Figure 1.   The exterior walls and appearance of the church of Bardewisch prior to restoration. Large patches of ancient works and plant roots penetrating the walls are characteristic of the entire building.

Figure 2. The church interior during restoration. Gothic paintings, which are in danger of being destroyed as are similar paintings on other walls, are visible to the left and far right. The middle section and especially the walls in the region of the window closed by bricks in later times (see Figure 1) are covered by large patches of algal and fungal infestations.

Figure 3.   Section from Figure 2 enlarged.   Fungal colonies
are clearly excluding algae at several places.   On the
other hand dark rims around the bright circles of fungal
colonies show that algae and fungi are in mutual relationship and interrelated in the sense of lichenized algal
growth pattern.

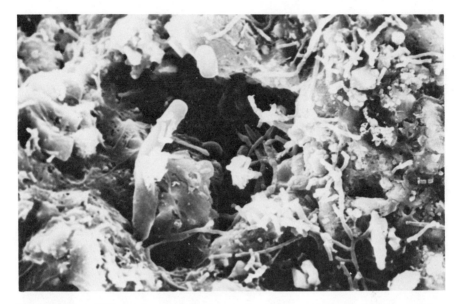

Figure 4. Filamentous blue-green algae, fungal hyphae and rod-shaped bacteria cover the mortar particles. Slime from decaying fungi and of the envelopes of cyanophytes is covering mineral particles in thin coatings. Scale: between bars 10 μm.

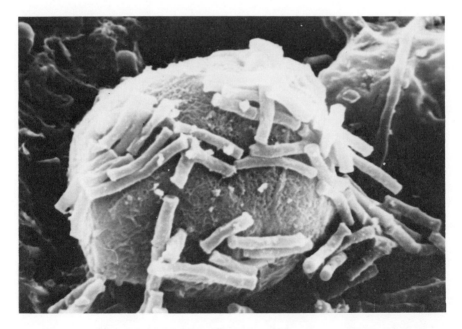

Figure 5. Coccoid blue-green algae covered with fungal hyphae. Upper right shows mucus cover of mineral phase. All SEM preparations were made by glutaraldehyde fixation, ethanol dehydration, replacement of ethanol by aldehydes and critical point drying. Scale: 3 μm between bars.

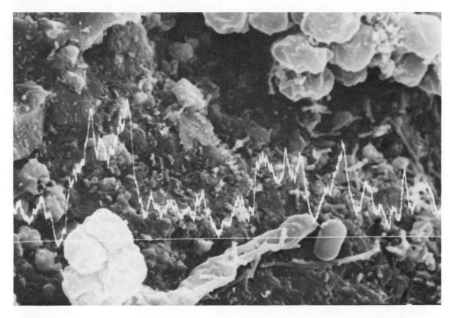

Figure 6.   Phosphorus and calcium linescan by EDX analysis.
Phosphorus (upper micrograph) is enriched and calcium low
in comparison to background of clay minerals covered with
fungal hyphae and bacteria.   Scale: 10 μm between bars.

Figure 7.   Phosphorus and calcium linescan by EDX analysis.
Phosphorus (upper micrograph) is enriched and calcium low
in comparison to background of clay minerals covered with
fungal hyphae and bacteria.   Scale: 10 μm between bars.

Heterotrophic bacteria, in succession to dying algae and fungi, are the main factors in mobilizing and transferring cations from the mineral phase to the liquid phase. This was shown by simple hot water extractions of surface parts infected by the microflora in comparison to permanently dry parts of the church walls, which expose only low populations and thus low mobilization rates.

The amount of cover on the surfaces of the walls was much higher than could be derived from the macroscopic view of the algal green patches. Fungal mycelia, which were invisible, surround the algal patches and occur at places that expose very little color change on the Gothic paintings themselves. Algal patches show only the centers of growth. Fungi and bacteria range much further and are even more active at some distance from the centers of growth. Fungal filaments and bacteria penetrate into the rock to depths of 3 to 6 mm.

These preliminary experiments were only the beginning of additional intensive studies of the phenomenon of growth and development of patch growth of lichenized algae and fungi in and on building stones and wall paintings.

DISCUSSION AND FURTHER ASPECTS

Building stones, plaster, wall paints and other wall covers such as plaster sculptures are populated heavily by fast growing fungi and algal populations in the damp and wet interior of the church of Bardewisch and at other places. Initial studies of such environments have yielded first approaches to the study of the associated detrimental effects of such growth.

The bacterial, fungal and algal flora associated with building materials and paintings are changing the appearance and esthetic qualities of buildings. Furthermore, elements are mobilized and the stability of cement and rock particles is reduced. In the Salla Terrena of Pommersfelden only the lower parts of the building have been infected, while in Bardewisch the whole church shows damage. Orientating studies by macroscopic, microscopic and cultural methods as well as by SEM and EDX analyses have shown that the flora is covering larger patches and penetrating deeper than can be seen in the church or residence itself.

Culture studies have demonstrated that the flora in many cases is lichenized and takes mutual advantage from light and humidity (algae) and the import of organic matter (fungi). The fungi are the main factors in penetration of rock particles and paint covers, and thus contribute largely to defoliation and other mechanical damages. Bacteria and fungi are the main factors in mobilization of cations but the cyanophytes are also enriching kalium and calcium by several magnitudes. It is planned to measure gas exchange rates in the field and in laboratory experiments and to quantify the influence of the microflora in comparison to sterile blank experiments, which measure exclusively the influence of humidity and temperature changes.

Weathering and destruction of building stones and mineral colors are largely influenced by the developing microflora when buildings and objects of art are not protected against humidity and light. In the case of Pommersfelden, humidity came from the underground with deviation of ground-water; isolation of the fundaments was proposed.

In Bardewisch it was necessary to take most of the plaster from the walls and dry out the church. Later a new roof and hydrophobic paints on the outside parts were added as additional protective measures. The paintings were isolated from the surroundings, dried separately, and will be washed with sterilizing agents. Many more studies are intended and needed to increase our understanding of the interrelated processes of biological and physical-chemical processes of rock decay.

REFERENCES

Aaronson, S. *Experimental Microbial Ecology* (New York: Academic Press, 1970), p. 236.

Berthelin, J. "Etude expérimentale de mécanismes d'altération des minéraux par des microorganismes hétérotrophes," *Ph. D. Thesis*, Nancy (1976).

Jaton, J. "Contribution à l'étude de l'altération micro-biologique des pierres de monuments en France," *Ph. D. Thesis*, Paris (1973).

Julien, A. A. "On the Decay of Building Stone, Part I and Part II," *New York Acad. Sci. II* (Trans.) 67-79, 120-138 (1882).

Kauffman, J. "Corrosion et protection des pierres calcaires des monuments," *Corr. et Anticorr.* 8(3):82-95 (1960).

Kessler, F. J. "Verwitterungserscheinungen von Bauwerken und Bausteinen," *Ph. D. Thesis*, Würzburg (1960).

Kieslinger, A. *Zerstörungen an Steinbauten* (Leipzig und Wien, 1932), p. 281.

Krumbein, W. E. "Zur Frage der Gesteinsverwitterung (über geochemische und mikrobiologische Bereiche der exogenen Dynamik)," *Ph. D. Thesis*, Würzburg (1966), p. 106.

Krumbein, W. E.   "Zur Frage der biologischen Verwitterung. Einfluß der Mikroflora auf die Bausteinverwitterung und ihre Abhängigkeit von edaphischen Faktoren," *Z. allg. Mikrobiol.* 8(2):107-117 (1968b).

Krumbein, W. E.   "Rôle des microorganismes dans la génèse, la diagénèse et la dégradation des roches en place," *Rev. Ecol. Biol. Sol,* IX:283-319 (1972).

Luckat, S.   "Luftverunreinigungen als Ursache des Steinzerfalls?" *Deutsche Kunst und Denkmalpflege* 31(1/2):45-50 (1973).

Marschner, H.   "Laboratoriumsuntersuchungen zum Verwitterungsschutz von Bausteinen," *Deutsche Kunst und Denkmalpflege* 31(1/2):32-44 (1973).

Pochon, J. and Y.-T. Tchan.   "Rôle des bactéries du cycle du soufre dans l'altération des pierres des monuments," *C. r. hebd. séanc. Sci.* (Paris) 226:2188-2189 (1948).

Riederer, J.   "Bibliographie der deutschsprachigen Literatur zur Verwitterung und Konservierung natürlicher Bausteine," *Deutsche Kunst und Denkmalpflege* 31(1/2):106-118 (1973).

Rosswall, T.   "Modern Methods in the Study of Microbial Ecology," *Bull. Ecol. Res. Comm.* (Stockholm), Vol. 17 (1973).

Weber, H.   "Ursachen und Behandlung der Steinverwitterung," *Maltechnik, Restauro* 2:73-89 (1977).

Winkler, E. M.   *Stone, Properties, Durability in Man's Environment* (New York: Springer, 1975).

CHAPTER 56

ENVIRONMENTAL ASSESSMENT OF SOILS IN CONTACT
WITH SULFUR-BASED CONSTRUCTION MATERIAL

E. J. LAISHLEY
M. G. TYLER

Department of Biology
The University of Calgary
Calgary, Alberta T2N 1N4, Canada

R. G. MCCREADY

Elliot Lake Laboratory
Canmet, E.M.R.
Ontario, Canada

INTRODUCTION

Western Canada and in particular the Province of
Alberta is the major center for elemental sulfur production
recovered from sour natural gas plants. New uses for the
stock-piled sulfur have been experimented with, especially
in the construction field using sulfur additions for producing
sulfur asphalt, sulfur loam and sulfur cements. (Sulphur
Institute Symposium, 1976). The physical wear and tear of
these construction materials has been assessed by visual and
mechanical testing to date. However, the environmental impact
on the soil ecosystem has virtually been ignored. It is well
known that sulfur can be oxidized in soils by sulfur-oxidizing
microorganisms, but the rate of this oxidization depends upon
particle size — the finer the sulfur particles the faster the
rate of oxidization to sulfuric acid. (Frederick and Starkey,
1948). This project was initiated to study the long-term
effects of soils in contact with sulfur-based construction
materials. We wish to report some of our preliminary findings
after two years on the chemical and microbial analysis
obtained from one of our four field test sites.

MATERIALS AND METHODS

Sulfur cement, portland cement and sulfur cylinders (6 in. dia x 12 in. length) were made and supplied by a group of civil engineers at The University of Calgary. The sulfur concrete contained 21 percent elemental sulfur and a sand-coarse gravel aggregate mix, which was identical to the portland cement, while the sulfur cylinders consisted of 100 percent elemental sulfur. Two of each type of cylinder were buried side by side with a 2.5-ft distance between the different types of cylinders in a silt loam field soil at the Canada Agricultural Research Station, Lethbridge, Alberta. Soil samples were removed at the times indicated in the tables, using a sterile metal core borer (1" dia) that was inserted adjacent to the side of the cylinders to a depth of 6." These soil samples (approx 50 g) were placed in sterile plastic bags and stored on ice while in transit to the laboratory. These soil samples were stored no longer than 48 hr at 5° C before microbial analysis was carried out.

The less acidophilic *Thiobacillus* counts were estimated by the most probable number (MPN) technique using a slightly modified Barton and Shively (1968) thiosulfate medium adjusted to pH 6.5 containing 50 µg/ml of bromothymol blue indicator. Five g of soil was added to 45 ml of sterile water in a 125-ml Erlenmeyer flask, mixed and adjusted to pH 6.5 with sterile 0.1 N H_2SO_4. From this solution, serial dilutions were made and 1 ml from the appropriate dilutions was added to tubes containing 5 ml of the above medium. These tubes were incubated at 28° C for 28 days; a positive tube was recorded when the indicator color changed to yellow (pH 6.0), indicating acid production plus confirmation by microscopic examination.

From these serial dilutions prepared above, the total number of bacteria were estimated by standard plate count on peptone yeast extract agar (Parkinson *et al.*, 1971) after incubation at 28° C for 7 days, while the sulfate reducers were determined by the method Mara and Williams (1970), after incubation at 28° C for 21 days. The acidophilic *Thiobacillus* counts were also estimated by MPN, with the exception of the thiosulphate medium, which was adjusted to pH 4.5 and contained 50 µg/ml bromophenol blue indicator.

Another soil dilution was prepared as described above but was adjusted to pH 4.5 before the appropriate serial dilutions were made. A positive tube was recorded when the indicator color changed to yellow, approximate pH 3.0, plus microscopic observation. A number of positive tubes from the different *Thiobacillus* counting media were streaked out onto thiosulphate noble agar medium adjusted to the appropriate pH, 6.5 or 4.5. Pure cultures of the less acidophilic *Thiobacillus* were restreaked twice before being subcultured in the less acidophilic *Thiobacillus* liquid thiosulfate medium. These organisms were identified by the classical methods outlined for this genus in Bergey's manual (1974). However, after the initial isolation of the acidophilic *Thiobacillus*, it was found that these isolates would not grow or would grow very poorly on subsequent subculturing on this thiosulphate noble agar and were not identified.

The pH of the soils was determined on a 1:1 ratio of soil to water using a Fisher accumet pH meter, and percent of organic matter was determined by the method of Davies (1974).

RESULTS AND DISCUSSION

The elemental sulfur cylinders were included in this investigation to obtain additional information on what may occur in soils surrounding piles of stored sulfur, a by-product of the sour gas plants.

The microbial assessment of the soil surrounding the sulfur, sulfur concrete and portland cement cylinders are shown in Tables I to III. The total count of heterotrophic

Table I
pH and Microbial Assessment of a Field Soil
Surrounding Buried Sulfur Cylinders at Lethbridge, Alberta

Sampling Time	% Organic Matter	pH of Soil	Number of Bacteria/g dry wt of soil	Number of *Thiobacillus*/g dry wt of soil	
				Less Acidic	Acidic
1974					
Oct.	4.9	7.7	$86 \times 10^5 (\pm 1.9)$[a]	Nil	Nil
1975					
May	6.6	7.0	$14 \times 10^5 (\pm 6.6)$	38	Nil
July	6.4	7.5	$19 \times 10^6 (\pm 3.8)$	96	Nil
Sept.	8.3	7.6	$66 \times 10^5 (\pm 3.0)$	2.8×10^2	Nil
1976					
May	4.0	7.4	$19 \times 10^7 (\pm 7.0)$	3.3×10^5	Nil
July	2.9	8.0	$29 \times 10^6 (\pm 5.0)$	1.3×10^4	296
Sept.	5.0	8.1	$29 \times 10^6 (\pm 7.0)$	1.6×10^4	5

[a]Standard error of the mean (SEM).

bacteria appears to be in the same order of magnitude in all test soils surrounding the various cylinders and showed normal fluctuation over the sampling period. The sulfur-oxidizing bacteria, due to their wide variation in pH tolerance for growth, were arbitrarily divided and estimated as two groups,

Table II
pH and Microbial Assessment of a Field Soil
Surrounding Buried Sulfur Cement Cylinders at Lethbridge, Alberta

Sampling Time	% Organic Matter	pH of Soil	Number of Bacteria/g dry wt of soil	Number of *Thiobacillus*/g dry wt of soil	
				Less Acidic	Acidic
1974					
Oct.	4.0	7.8	$26 \times 10^8 (\pm 1.4)^a$	Nil	Nil
1975					
May	6.4	7.3	$19 \times 10^5 (\pm 3.0)$	62	Nil
July	5.6	7.5	$19 \times 10^6 (\pm 3.8)$	96	Nil
Sept.	5.9	7.6	$66 \times 10^5 (\pm 3.0)$	2.8×10^2	Nil
1976					
May	3.0	7.2	$96 \times 10^6 (\pm 7.0)$	2.5×10^4	126
July	3.4	8.0	$41 \times 10^6 (\pm 2.0)$	1.1×10^3	Nil
Sept.	4.9	8.1	$27 \times 10^6 (\pm 8.0)$	5.1×10^4	131

[a]SEM.

Table III
pH and Microbial Assessment of a Field Soil
Surrounding Buried Portland Cement Cylinders
at Lethbridge, Alberta

Sampling Time	% Organic Matter	pH of Soil	Number of Bacteria/g dry wt of soil	Number of *Thiobacillus*/g dry wt of soil	
				Less Acidic	Acidic
1974					
Oct	4.3	7.7	$72 \times 10^5 (\pm 0.4)^a$	Nil	Nil
1975					
May	5.8	7.6	$80 \times 10^5 (\pm 3.8)$	Nil	Nil
July	6.0	7.5	$15 \times 10^6 (\pm 5.1)$	Nil	Nil
Sept.	8.0	7.8	$83 \times 10^5 (\pm 1.6)$	19	Nil
1976					
May	5.9	7.4	$12 \times 10^7 (\pm 1.0)$	162	Nil
July	2.4	8.5	$68 \times 10^6 (\pm 1.0)$	Nil	Nil
Sept.	4.7	8.3	$35 \times 10^6 (\pm 1.0)$	Nil	Nil

[a]SEM.

the less acidophilic (*T. thioparus* group) and the acidophilic (*T. thiooxidans* group). This was done to see if we could detect a succession in sulfur oxidizers from the less acidophilic to the acidophilic species in soils surrounding the different sulfur cylinders as was shown by Parker (1947) in the corrosion of concrete sewer pipes. The sulfur oxidizers were not detected at the zero time sampling in October 1974, but the less acidophilic *Thiobacillus* were detected in soils surrounding the sulfur and sulfur cement cylinders in the May and July 1975 sampling. However, their numbers increased significantly in these soils by the September sampling in both cases. The less acidophilic sulfur oxidizers were only detected in the September samplings around the Portland cement cylinders. Large populations of the less acidophilic *Thiobacillus* were detected throughout the 1976 sampling period in soils surrounding the sulfur and sulfur cement cylinders, while the acidophilic *Thiobacillus* were being detected around the different sulfur cylinders only during the 1976 sampling.

In contrast, no acidophilic oxidizers were detected in soils surrounding the Portland cement cylinders, while the less acidophilic *Thiobacillus* were only detected in the May 1976 sampling. This clearly shows that the different sulfur cylinders were selecting for the sulfur-oxidizing bacteria. There was no consistent decrease in soil pH occurring around the different sulfur cylinders even though significant populations of sulfur oxidizers were found. This was probably due in part to the buffering capacity of the soil. In fact, the soil pH varied from near neutrality in the May sampling and increased significantly in alkalinity by the July through September sampling, especially around the sulfur and sulfur cement cylinders. It cannot be ascertained at this time whether this trend will continue. It should be noted here that when we repacked the soil around the cylinders at the start of the study in 1974; these soils contained a mixture of the different horizons Bm, Cca and Ck ranging in pH from 7.0 to 8.1 (Canada Department of Agriculture data), which could explain in part some of the fluctuations in the pH readings obtained so far. This mixture of soil horizons was also indicated by the fluctuations observed in the percent organic matter determined in these soil samples at the different sampling times.

We have isolated and identified a number of less acidophilic *Thiobacillus* sp. as follows: *T. intermedius, T. perometabolis, T. thioparus* and some unidentified *Thiobacillus* sp. along with a number of *Pseudomonas* sp. from our counting medium. These *Pseudomonas* sp. were found to oxidize thiosulfate, which was the sulfur source in the counting medium to tetrathionate and an unidentified polythionate. Even though these *Pseudomonas* could sometimes be detected in our counting medium, they did not produce any

acids on oxidation of thiosulfate.   Thus they did not inter-
fere with the *Thiobacillus* counts, which depended on acid
production for recording a positive tube in the MPN counts.
      We also tried to estimate the sulfate-reducing bacteria
in these soils, but upon isolating some positive colonies
from the selective roll tube counting media we found not
only the classical *Desulfovibrio* organisms but some *Bacillus*
sp.   Further examination of these *Bacillus* sp. showed they
could reduce both sulfate or sulfite to hydrogen sulfide.
Consequently, false positive counts were obtained, and the
true sulfate reducers could not be differentiated from these
Bacillus in the selection medium by colony morphology.   Thus
we could not use these data.
      Considerable fluctuation in soluble sulfate made
correlation of the elemental sulfur and total sulfur values
determined in these soils with the sulfur oxidizer counts or
pH determinations an uncertainty.   To date we have found that
at this site and at the other three sites when solid sulfur-
containing materials are placed in the soil, significant
populations of a less acidophilic *Thiobacillus* were found
within a year and those populations were maintained throughout the second year.   However, no drastic drop in soil pH
has occurred; this will probably depend upon the buffering
capacity of each particular soil.   The acidophilic *Thiobacillus*
have recently been detected at this site and in a forest
soil near Battle Lake, Alberta, in soil surrounding the sulfur
and sulfur cement during the September 1976 sampling.

ACKNOWLEDGMENTS

      This investigation was supported by The University
of Calgary Interdisciplinary Sulphur Research Group (UNISUL).
The authors also acknowledge the technical assistance of
Miss E. Bigornia and Mrs. T. Lange.

REFERENCES

Barton, L. L. and J. M. Shively.   "Thiosulfate Utilization
   by *Thiobacillus thiooxidans*," ATCC 8085. *J. Bacteriol.*
   95:720 (1968).

Buchanan, R. E. and N. E. Gibbons, Eds. *Bergey's Manual of
   Determinative Bacteriology*, 8th ed. (Baltimore: Williams
   and Wilkins, Co., 1974), pp. 456-461.

Davies, B. E.   "Loss on Ignition as an Estimate of Soil
   Organic Matter," *Soil Sci. Soc. Amer. Proc.* 38:150-151
   (1974).

Frederick, R. L. and R. L. Starkey. "Bacterial Oxidation of Sulfur in Pipe Sealing Mixtures," *J. Amer. Water Works Assoc.* 40:729-736 (1948).

Mara, D. D. and D. J. A. Williams, "The Evaluation of Media used to Enumerate Sulphate Reducing Bacteria," *J. Appl. Bact.* 33:543-552 (1970).

New Uses for Sulphur and Pyrites, The Sulphur Institute Madrid Symposium (London: Lynwood House, 1976), pp. 1-236.

Parker, C. D. "Species of Sulphur Bacteria Associated with Corrosion of Concrete," *Nature* 159:439-440 (1947).

Parkinson, D., T. R. G. Gray and S. T. Williams. "Media for Isolation of Soil Micro-Organisms," in *IBP Handbook No. 19 Methods for Studying the Ecology of Soil Micro-Organisms* (Oxford and Edinburgh: Blackwell Scientific Pub., 1971), pp. 105-109.